Trailside
Botany

101 Favorite

Trees, Shrubs, & Wildflowers

of the Upper Midwest

Trailside Botany

Botany

101 Favorite

Trees, Shrubs, & Wildflowers

of the Upper Midwest

Written by John Bates
Illustrated by April Lehman

Pfeifer-Hamilton
Duluth, Minnesota

Pfeifer-Hamilton Publishers
210 West Michigan
Duluth, MN 55802-1908   218-727-0500

*Trailside Botany: 101 Favorite Trees, Shrubs,
& Wildflowers of the Upper Midwest*

Map on page x is used with permission by *A Sierra Club
Naturalist's Guide to the North Woods* © 1981 Sierra Club
Books.

Printed in the United States of America by Edwards Brothers
Incorporated
10 9 8 7 6 5 4 3 2 1

♻ Printed on 50% total recovered fiber, 10% post-consumer fiber

Editorial Director: Susan Gustafson
Graphic Design: Jeff Brownell
Cover Photo: Paul Hapy

Library of Congress Cataloging in Publication Data

Bates, John, 1951–
    Trailside botany : 101 favorite trees, ferns, & wildflowers of
the upper Midwest / written by John Bates ; illustrated by
April Lehman.
        p.   cm .
    ISBN 1-57025-070-7 (softcover)
    1. Botany—Middle West—Identification. 2. Trees—
Middle West—Identification.   3. Wild flowers—Middle West—
Identification.   4. Botany—Great Lakes Region—Identification.
5. Trees—Great Lakes Region—Identification. 6. Wild
flowers—Great Lakes Region—Identification. I. Lehman, April.
II. Title.
QK128.B38   1995
581.977—dc20

                                    95-14380

For Mary

Special thanks go to Dr. Jim Meeker for his review of the manuscript and many suggestions; and to Dr. Keith White, Dr. John Reed, Dr. Richard Presnell, and Gary Fewless who all greatly influenced my understanding of the plant world.

Thanks are also due to Mary Shafer for her preliminary drawings that helped serve as a basis for the final illustrations and to April Lehman, whose artistic talent and love of plants shines through in the illustrations.

Additional thanks must go to the staff at Pfeifer-Hamilton for publishing regional natural history books that help those who love the North Woods to better understand their environment.

My greatest thanks go to my wife Mary, whose enthusiam and love for the natural world continues to inspire me every day.

Table of Contents

Introduction ix

Deciduous Trees

Sugar Maple 2
Red Maple 6
Silver Maple 8
Northern Red Oak 10
Quaking Aspen/Big-tooth Aspen 13
Paper Birch 16
Yellow Birch 19
Basswood 22
Cherry—Black, Pin, Choke 25
Ironwood 27
Black Ash 30

Conifers

Eastern Hemlock 34
Balsam Fir 37
White Pine 39
Red Pine 42
Jack Pine 45
Black Spruce 47
White Spruce 49
White Cedar 51
Tamarack 53

Shrubs

Staghorn Sumac 56
Blackberries/Red Raspberries 58
Red Osier Dogwood 61
Beaked Hazelnut 63
Highbush Cranberry 65
Willow 68
Tag or Speckled Alder 71
Blueberry 73
Sweet Gale 75

Maple-leaved Viburnum 77
Thimbleberry 79
Sweet Fern 81

Early Spring Wildflowers

Trailing Arbutus 84
Wood Anemone 86
Marsh Marigold 88
Blue-bead Lily 90
Hepatica 92
Gaywing 94
Spring Beauty 96
Bloodroot 98
Trout Lily 101
Sessile-leaved Bellwort 103

Late Spring Wildflowers

Barren Strawberry 106
Wild Strawberry 108
Oxeye Daisy 110
Starflower 112
Large-flowered Trillium 114
Bunchberry 117
Columbine 119
Solomon's Seal 121
Canada Mayflower 125
Goldthread 127
Orchids 129

Early Summer Wildflowers

Orange Hawkweed 134
Common Milkweed 136
Wintergreen 139
Sarsaparilla 141
Black-eyed Susan 144
Indian Pipe 146
Pipsissewa 148
Twinflower 151

Late Summer Wildflowers

Goldenrod	154
Large-leaf Asters	156
Joe-pye Weed	158
Fireweed	161

Bog Plants

Sphagnum Moss	164
Labrador Tea	169
Cottongrass	171
Bog Laurel	173
Bog Rosemary	175
Leatherleaf	177
Sundew	179
Small Cranberry	181
Pitcher Plant	183
Bladderworts	185

Wetland/Open Water Plants

Water Lilies	188
Common Cattail	190
Purple Loosestrife	192
Pickerelweed	194
Blue Flag Iris	196
Horsetails	198
Arrowhead	201
Wild Calla	203

Other Plants

Clubmosses	206
Lichens	209
Ferns	213
Bracken Fern	217

Glossary/Sources

Glossary	220
Sources	224

Introduction

Four summers ago a visitor to the North Woods who had gone on several hikes with me said, "Whenever I hike in the woods I get frustrated because I don't know what I'm seeing. What I need is a book about the most common species, how to identify them, and what I should know about them." He wanted a book of the one hundred most common plants and animals of the North Woods so he wouldn't have to carry a whole bundle of separate bird, flower, and mammal guides along on his trips. Not only didn't he want to carry all that weight, but those guides all include far more species than he cared to know about, and many more species than he would ever see in the North Woods.

What's more, he wanted me to write the book.

So I dove in. I soon discovered the impossibility of limiting myself to fifty species of plants and fifty species of animals. I decided to focus just on plants, and I selected the hundred and one species common to the experience of most North Woods hikers.

All the species described in this book are found in the North Woods, an area of transitional forest that extends from north central Minnesota to Maine, and up into southern Canada. No definitive border separates the boreal forest to the north, the southern forest, and the North Woods in between. They simply grade into one another. However, within the North Woods area many southern species reach their northernmost range, and many northern species reach their southernmost range.

The North Woods area of Minnesota, Wiisconsin, and Michigan

The map above shows the extent of the North Woods in Minnesota, Wisconsin, and Upper Michigan. The border, of course, is not as definite as the map might suggest.

Trailside Botany is not intended primarily as an identification guide, though I do note key characteristics to look for in each species and the means by which to remember them. This book offers something more—a detailing of what is interesting and important about each specific plant.

Every plant has a story—often just as interesting as those told about the "glamor" wildlife species like wolves and whales. William Least Heat-Moon writes, "Even stationary objects become travelers as they move far through time, gathering their legends along the way, their tales publicly told or privately imagined, tales to be shared like communion wine." Here I have tried to write each plant's "story," a sort of thumbnail biography of its natural and cultural history.

This book encourages you to observe carefully and see, not just look. To truly "see" what is taking place in a given site, you must understand some of the processes, interrelationships, and evolutionary adaptations that allow our northern flora to withstand the rigors of five to six months of snow, cold, and ice.

The observations and activities that I suggest will enhance your understanding, but exercise caution in any experiment that might harm a plant. Note first whether a community of plants exists. Only if the species is numerous, should you proceed with an activity that might disturb it. I was once taken to task by a hiker who believed it to be absolutely wrong ever to pick a flower or leaf. My response to her was that if we estrange ourselves from any physical contact or wise experimentation with the natural world, we will always remain apart from it. If you know an area well, understand the natural history of a plant, and recognize that no harm will be done to a community of plants by collecting one individual, then I believe you can responsibly proceed with your investigation.

The glossary, which begins on page 219, defines botanical terms that may be unfamiliar. Among the first characteristics to note when identifying a plant are whether the leaves are simple or compound, pinnate or palmate, and whether they are alternate or opposite on twigs. Check the glossary for clarification of terms such as these.

I use the Latin and scientific terminology in this book as little as possible, but Latin names do perform the essential task of eliminating confusion created by the profusion of common names for any one species. (For instance, bracken fern is also known as brake, pasture brake, adder's spit, eagle fern, oak fern, umbrella fern, upland fern, and turkey foot bracken.) Pay attention to the Latin root words. The Latin genus and species names often illustrate some function or feature of the plant that will help you remember it.

I refer often to historical uses in my descriptions of each species. Remember that the medicinal merits of any given plant were based on equal parts of fact, fiction, fantasy, and falsehood. Toss in for good measure pinches of mysticism, necromancy, mythology, alchemy, astrology, and basic magic, along with honest efforts at good botany and pharmacology. You will have to make your own best judgments on whether these historical uses have any merit today, but I suggest that you be cautious about ingesting any wild plant.

I invite you to use this book as a starting point for your exploration of the North Woods. When you observe plants carefully, you gain a new understanding of their natural history and your outdoor experience is greatly enriched.

Deciduous Trees

Sugar Maple

Red Maple

Silver Maple

Northern Red Oak

Quaking Aspen/Big-tooth Aspen

Paper Birch

Yellow Birch

Basswood

Cherries—Black, Pin, Choke

Ironwood

Black Ash

Sugar Maple

Acer saccarim

Maple seedlings cover the ground in this mature maple and hemlock forest.

Sugar bush

Smooth five-lobed leaves

Sugar maple leaves usually have five lobes, with the edges of the leaf being smooth or having a few wavy teeth. Red and silver maple, the North Woods' other two native maple trees, both have leaves with serrated edges, so are easily distinguished from sugar maples.

The Native Americans were the first to boil down the nearly tasteless sap. Every Ojibwa family or group had its own sugar bush, to which it moved in mid-March for a month of sugaring. Frances Densmore, who lived with the Ojibwa for many years in the early 1900s, wrote that an average-size camp drilled nine hundred taps, though the larger camps used up to two thousand taps.

A normal camp would produce a hundred or more gallons of syrup every spring.

It took a long time to cook down all the sap (forty gallons of sap makes one gallon of syrup), and the syrup was not the end stage. "Sugaring off" took place later. The syrup was further cooked down until it thickened, and then was worked into a granulated sugar. The Ojibwa filled birchbark vessels, called makuks, with the sugar and stored it for use throughout the year. They used this sugar as a seasoning for virtually everything, from fruits to fish. Dissolved in water, the sugar became a cooling summer drink (the first soft drink?). Mixed as a syrup base for medicines, it made the often-bitter remedies palatable. Cones of birchbark filled with syrup, allowed to harden, became gifts or treats for children.

Aside from its value as a sweetener, sugar maple plays a dominant ecological role in the forests of the North. No other species in the North Woods can withstand living in the shade as long and well as the sugar maple. Because of this ability to withstand "suppression," sugar maple seedlings less than 1/2 inch in diameter are often more than forty years old.

Sugar maples produce enormous quantities of seeds. A twelve-year study of one site found that the number ranged from 40,000 per acre to over 5 million per acre. (An acre is 208 feet on a side, or about two-thirds the length of a football field.) Maple seeds, the winged helicopters many of us played with as children, are shed at the same time as the leaves fall, and so receive the advantage of being buried in the nutrient-rich leaf litter. As a result, huge numbers of young seedlings occupy the forest floor, commonly 20,000 per acre that are less than 12 inches tall. One researcher estimated that 4,000 young seedlings spring up every year under one mature sugar maple, totaling over one million seedlings in the life of a 350-year-old tree. Out of those only about 5 percent (about 50,000) last into the second year; maybe 1,400 live to 10 years; perhaps 35 grow over 20 feet tall; two may reach 150 years old; and one may be lucky enough to reach maturity.

Once a sugar maple finds a gap in the canopy and races to the sky, it is likely to live for well over two hundred years. Toss in one last longevity factor, the ability to withstand for many years being browsed by wintering deer (a characteristic hemlock doesn't share), and sugar maple often becomes the dominant tree in a rich northern hardwood forest.

So why aren't they everywhere? Simply because we tend to cut our forests so frequently that those species which love sun are favored over those that can tolerate shade. The pioneers (aspen, white birch, cherry, and jack pine) can outcompete sugar maple on a wide open site. But if that young forest is allowed to grow for several hundred years, sugar maple will begin to take over.

One last factor helps sugar maple dominate in mature forests. Leaves dropped from sugar maples in the fall are rich in nutrients and significantly enrich the soil. Of the thousands of seeds that germinate in this fully shaded, high-nutrient soil, enough reach the sky for sugar maples to ensure their reign in the older forests of the North Woods.

ACTIVITY

Measure how many seedling sugar maples are biding their time in hopes of piercing the canopy. Build a simple quadrat, a fancy term for a 2' x 2' square of wood, by nailing four 2-foot-long 1' x 2's together into a square. In a mature hardwood forest toss the quadrat into a "sea" of sugar maple seedlings. Count the number of seedlings (look for the five-lobed, smooth-edged leaves) inside the 4-square-foot enclosure. Since an acre is about 208 feet on a side, or 43,560 square feet, multiply your seedling number by 11,000 (10,890 if you want to be more exact) to find the number of seedlings per acre. An acre of mature trees might contain only one hundred to two hundred trees, so consider the vast number of seedlings that won't ever make it.

Red Maple

Acer rubrum

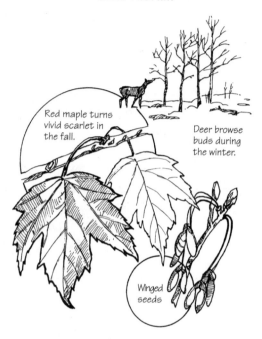

Red maple turns vivid scarlet in the fall.

Deer browse buds during the winter.

Winged seeds

R ed maple leaves have three to five lobes with teeth, or serrations, all along the edges. A visual aid may help you remember that sugar maple is smooth along the edges, while red maple is jagged: see those serrated edges of the red maple as being able to cut you and draw blood. Red blood, red maple.

Red maple, a "soft" maple, produces fewer BTU's when burned and is weaker than sugar maple (a "hard" maple), and thus is far less prized. The red maple also differs in that it seldom reaches great age or large size, and it doesn't achieve the dominance that sugar maples do in mature forests. But it beats sugar maple in its ability to grow in both wet and dry soils and in both sunny young forests or shady old forests. Red maples can be found just about anywhere in the North Woods, being possibly the tree most tolerant of varying habitats there.

Red maples stand out in particular in the fall, when they turn a vivid scarlet. Neither silver nor sugar maple can match the brilliance of the red maple in autumn.

Deer are happy to browse red maples, and chipmunks, squirrels, and a host of rodents dine on the winged helicopter seeds that drop in the spring. The red maple's sap can be used for syrup, but is less favored than that of the sugar maple.

ACTIVITY

Sometimes in the autumn red maple leaves turn both a brilliant red and yellow due to uneven sunlight on the leaves. Red pigments require sunlight to develop, while yellow pigments develop without sunlight. To prove this, cover a portion of a red maple leaf with opaque tape in the late summer. In the fall, remove the tape after the rest of the leaf has turned red—the "shaded" portion of the leaf should be yellow.

Silver Maple

Acer saccharinum

Silver maples can be found along streams and in flood plains.

The leaf of the silver maple has five lobes, each lobe indented or notched more than halfway to the center of the leaf. The leaf edges are serrated and the leaf is silvery white on the underside. Neither the red nor the sugar maple has deeply notched lobes like those of the silver maple.

Silver maple, a lowland species, commonly grows along stream banks and in floodplains. Many occur along the Manitowish River just below my house, and one big old fellow stands alone in our wetland, which is often flooded in the spring. This is a tough place to survive if you're not adapted to periodic submergence.

You won't find silver maple in your local lumberyard—the wood is brittle and decays rather easily. And syrup tappers don't get enough sap from silver maple to warrant the work. Some say the twigs even smell bad when broken.

The trunks can get quite large in diameter, and the hearts often rot, providing excellent dens for small mammals like raccoons and squirrels, and for cavity-nesting birds like wood ducks, great horned owls, pileated woodpeckers, and hooded mergansers. I often see "hoodies" taking cover under the low-hanging branches of silver maples on the Manitowish River.

The large seeds feed rodents (mainly squirrels) and many birds, especially pine and evening grosbeaks.

OBSERVATION

Silver maples take advantage of their usual spring swim by dropping their seeds in late spring, most often on the freshly opened soils that the shrinking flood waters leave behind. Note when the seeds ripen (usually in May), and see if that time period coincides with the usual receding of flood waters in your area.

Northern Red Oak

Quercus borealis

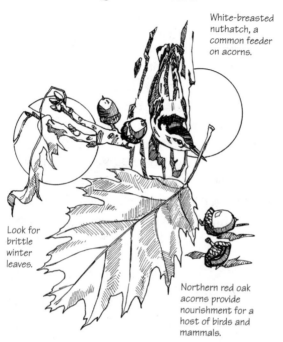

White-breasted nuthatch, a common feeder on acorns.

Look for brittle winter leaves.

Northern red oak acorns provide nourishment for a host of birds and mammals.

R ed oak is the only oak of tree size whose range extends well into the North Woods. It has seven to eleven lobes on a leaf, each with a pinpoint at the tip (white oaks are rounded at the lobe tips). The acorns are large, 3/4 to 1 inch long, with a saucer-shaped "hat." The winter twigs have a cluster of pointed buds at their tips.

Acorns provide an important part of the wildlife diet, particularly in the winter. One study showed that acorns made up 62 percent of a Wisconsin wood duck's diet. North Woods species like ruffed grouse, mallards, flickers, grackles, blue jays, nuthatches, brown thrashers, varied thrushes, redheaded woodpeckers, black bears, chipmunks, squirrels, raccoons, mice, and deer are just some of the animals who seek out the red oaks' acorns.

But to humans, acorns are bitter and inedible unless the strong but soluble tannic acids have been leached out by flowing water. American Indians did just that, and then pounded the kernels into a meal.

Where limbs fall or are broken off, woodpeckers hollow out holes in their search for insects, providing cavities for nests of owls, squirrels, chickadees, and a few less desirable inhabitants, like starlings.

Red oaks are better adapted to low-light levels than most oaks, being able to grow in the sun-flecked shade of aspens and birches. They thrive best on rich, well-drained, and moist soils, and can live several centuries, fattening up to 2 to 3 feet in diameter.

Fire usually kills a red oak, because the bark is susceptible to damage and the tree rarely sprouts again after a fire. But if you cut down a red oak, the stump will send forth a host of sprouts to compete for the new opening in the leaf canopy. Try to grub the stump out, and you face a taproot of great strength. Transplanting saplings doesn't work too well because of the taproot.

Red oak's heavy, strong wood is used as flooring, in furniture (though not the best, because of the wood's open pore structure), for posts, and for firewood.

ACTIVITY

Gather acorns in September. Fill a clear glass container with soil and plant the acorns close enough to the edge so you can examine the young tap root as it grows. Transplant the seedlings the next spring onto a site with some shade and some sun. Generations to come will thank you, as will the wildlife that will utilize the acorns, the cover, and the nesting and denning opportunities that your tree provides.

Quaking Aspen / Big-tooth Aspen

Populus tremuloides
Populus grandidentata

B oth quaking aspen and big tooth aspen have simple,
alternate, broadly oval leaves with flattened stems,
which allow them to tremble, or "quake," in the wind.
The rustling sound of the leaves in a wind provides an
identifying feature itself. Quaking aspen leaves have
small teeth, while big-tooth aspen leaves have, not surprisingly, big teeth.

These two aspens, or popples as they are also
known, are the most abundant trees in the North
Woods. They composed only 1 percent of the
presettlement forest, but now make up 26 percent of
northern forests. Their extraordinary increase followed
the logging of the North Woods. Aspens love full sun
and were "waiting in the wings" for such a massive
forest clearing.

Following clear-cut logging, aspen invades the open ground in a number of ways, the most remarkable being root suckering. Aspens send out root suckers, which turn upward and surface as the beginning of new trees. These clones are literally holding hands underground. Take a shovel sometime and dig between two aspens in a young aspen stand to see this connection. One clonal group in Colorado occupies over 200 acres, with one genetic individual represented by more than 47,000 trees. Another clonal group in Minnesota, covering several acres, is estimated to date back 8,000 years to the last glaciation.

Aspen suckers number up to 30,000 or more per acre. These suckers create what's called a "dog hair" stand because they grow so thick they're like hair on the back of a dog. Such a woods is just about impossible to walk through. But as the stand grows up— and it grows fast, about 3 feet per year—competition naturally thins out the "hairs" until 400 or fewer trees per acre are left at maturity. The process takes about forty years for an aspen stand. Foresters usually come in at this point and clear-cut the aspen. Then the process starts all over again. Left alone, aspens usually die by eighty years of age.

Root suckering alone, however, couldn't account for the aspens' rise to dominance. Instead, they took over the current North Woods by also being prolific seed producers. On breezy spring days the white silky seeds come down like a blizzard, and are carried away on the wind. The tiny seeds travel just about everywhere, so wherever a forest opening occurs, aspen seeds will probably find it.

The Ojibwa women used a root decoction to help

stem excessive menstrual flows, and white settlers used the bark for a quinine substitute. People today mainly value aspen as a pulpwood—most logging trucks leaving the North Woods carry aspen bound for the paper mills.

Aspen has great wildlife value too. Grouse eat male aspen buds as the main component of their winter diet. Deer browse the young twigs and foliage. Beaver and porcupines love the soft inner bark. Snowshoe hares devour the twigs and young bark. The moose population prizes the aspen as a winter browse. Butterflies such as the purple banded, the tiger swallowtail, and the viceroy eat the leaves.

ACTIVITY

Pick an aspen leaf and look closely at how the stem is flattened. The planed surface can catch the wind and twist back and forth, "quaking" and trembling with a sound like raindrops. William Least Heat-Moon writes that the sound of cottonwoods (a close relative of the aspens) in the wind is "a liquid sound that could almost quench a thirst."

Paper Birch

Betula papyrifera

Birch pot
for cooking

No other tree in the North Woods has white, papery, shredding bark with long, blackish, horizontal lenticels (breathing pores). The simple, alternate leaves with double-toothed margins look like the leaves of many other trees and thus pose a more difficult identification task. The young saplings bear little resemblance to the adults, possessing smooth, shiny, reddish-brown bark that looks like the bark of a cherry tree.

White birch is a pioneer species. Like aspen, it loves sunlight, and is one of the first trees to invade a newly burned or cutover forest. Once it's established, birch is hard to kill. Stumps often sprout new shoots, and those that survive grow in the clumps of three to six stems that are so common in the North Woods.

White birch acts as a "nurse" tree for ensuing generations of trees that grow underneath it and can tolerate shade, such as red maple, white pine, balsam fir, elm, and red oak, but white birch can't survive in its own shade. It grows fast, rarely lives for more than ninety years, and decays quickly after it falls.

The white birch's survival strength lies in its ability to outcompete other species on fire-scarred soil. The thin, quickly ignited bark fuels fires that destroy the evergreens and shade-tolerant hardwoods in whose shade the birch can't survive. While the birch too dies in the fire, its seeds will be the first to colonize the area, and another generation will be assured.

But white birch is on the decline in the North Woods. Today forest managers do such a good job at preventing fires that white birch seldom gets the opportunity to make use of its fire-adaptation skills. Add the devastating effects of the bronze leaf borer and the birch leaf miner that swept through the North Woods after the droughts in the late 1980s, and white birch appears destined for a much smaller role in the immediate future.

Historically, white birch played an indispensable role in Native American and French voyageur society. In fact, an argument can be made that the history of North America would be altogether different if it were not for the white birch. The birch provided the skin for canoes that carried the first Europeans on economic, religious, and political expeditions into the wilderness. No other material was nearly as light, durable, waterproof, or strong as birch bark.

Native Americans used the bark for waterproof sheeting over wigwams, for pots to cook in (birch pots could be placed directly over a fire if the inner bark surface was exposed to the fire), for boxes in which to cache food, and for just about every kitchen utensil they needed. Other uses included tinder, torches, fans, patterns for beadwork, a medicine for stomach pain, transparencies and etchings for artwork, cones for maple sugar treats or for handy temporary containers, and even wrapping for the bodies of the dead.

Twelve bird species, including grouse, redpolls, and pine siskins, eat white birch seeds. Deer, moose, and snowshoe hares browse the buds and twigs and the inner bark provides a treat for beavers and porcupines.

OBSERVATION
Look at the understory of a pure stand of white birch—few, if any, white birch seedlings will be present since they can't tolerate shade.

ACTIVITY
Take a piece of birch bark from the ground (not peeled off a live tree), soak it in water, and put a match to it. It will spit a bit, but it will light up just fine, as any drenched hiker knows who has tried to start a campfire in the rain.

Yellow Birch

Betula lutea

Yellow birch
"standing on
its toes"

Yellow birch has alternate, simple, double-toothed leaves. (These descriptive terms, which are used throughout the book, are defined in the glossary.) The distinguishing characteristic is the yellow-bronze, shiny, peeling bark. It looks similar to white paper birch bark, but is obviously yellow instead. Saplings provide more of a challenge to identify since the yellow bark has yet to develop. Break off the end of a new twig and chew on it—if it tastes like wintergreen, it's yellow birch. The bark was once commercially harvested to produce wintergreen oil.

About the only qualities yellow and white birch

have in common are their peeling bark and their flower and seed structures. Otherwise they are ecologically worlds apart. Yellow birch is a climax species, meaning it tolerates shade very well and grows in mature forests. Common associates are sugar maple, hemlock, and basswood, all trees of old-growth stands. Unfortunately, yellow birch is relatively uncommon because we don't allow most of our forests to mature. Yellow birch grows slowly, gets big (2- to 4-foot diameters are common), and lives a long time (150-300 years).

The two-winged seeds drop in late autumn and are a common sight in the winter, skittering along ski trails, seeking a new germination spot. Yellow birch germinate frequently on old stumps or tip-up mounds, where their tiny winged seeds can get a foothold more easily than they can through the thick leaf litter on the forest floor. Over time the stump rots away, and many a yellow birch has been left "standing on its toes," like a bronze ballet dancer frozen in space. Some of these arching roots are spread so far apart that a person can crawl between them. The wind often fells these vulnerable trees.

Yellow birch becomes a cavity nesting tree deluxe. The hearts of the big old trunks often rot away, providing denning sites for large mammals like black bears, raccoons, porcupines, and fishers. In their sapling stage, yellow birch make a fine winter browse for whitetail deer and moose. Redpolls, chickadees, and pine siskens, among others, eat the seeds.

Woodworkers scour the forests, looking for yellow birch with burls (growths from bacterial diseases, protruding like massive warts on the trunks), which they make into exotic tabletops and woodenware. The burl wood grain runs in every direction, creating a unique pattern each time.

OBSERVATION

Look for yellow birch reproduction on old rotting stumps and mossy "nurse" logs. Once the stump or log rots from beneath them, these saplings will be birch "ballet" dancers if wind or snow doesn't throw them down first.

Basswood

Tilia americana

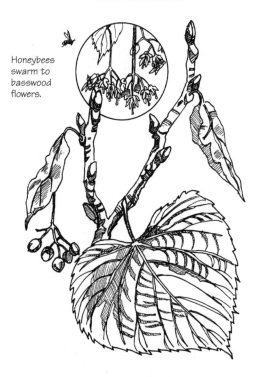

Honeybees swarm to basswood flowers.

The alternate, simple, sharply toothed leaves are highly distinctive due to their large size (5 to 6 inches long and 3 to 4 inches wide) and their asymmetrical, heart-shaped base.

Basswoods often grow in perfect circular clumps due to their ability to "crown sprout." When the central trunk dies, sprouts grow at the base of the trunk and develop into mature trees. Each of these trees then creates root crown sprouts too, so when they die the shoots grow and the circle of basswoods is perpetuated. This vegetative reproduction strategy keeps basswood competitive in the old-growth forests so prone to being overrun by sugar maple.

Basswood tolerates shade just fine, and is a climax species. It grows after the pioneer species (usually aspen and white birch) have died, and after their successors, such as red maple, red oak, white pine, have died too or have become somewhat older and more sparse.

The flowers are basswood's highlight. Honeybees swarm to them and create distinctly flavored basswood honey. A fine tea can be brewed from the flowers.

The seeds form in a group of gray pea-sized hard nutlets dangling from a structure called a bract. Squirrels, chipmunks, and rabbits eat these fruits but the fruits aren't a wildlife mainstay. Deer browse the young twigs, and as with most tree species, porcupines dine on the inner bark.

Historically, basswood was one of the most important materials in the life of the Ojibwa. They made twine from the layer of fiber, the cambium, just under the bark. They cut sheets of bark from as high as a person's reach down to the ground, and then cut those into strips about 4 inches wide. They soaked the strips in water along a lake edge for about ten days, after which the inner yellow fiber would easily detach. Lengths of fiber were then cut for different purposes: wide strips for making pitch bags for boiling gum to repair canoes, thinner strips for making woven bags, and very thin strips for making twisted twine. The Ojibwa used the twine to tie the poles of wigwams, to bind sheaves of wild rice, to fasten stones on fish nets as sinkers, to make floor mats, and for a multitude of various sewing projects.

Basswood

Observation

Look for a circular clump of basswood, a charac-teristic growth form of basswood after it's been cut. From the diameter of the circle in the center of the clump, estimate the width of the original central trunk.

Cherries—Black, Pin, Choke

Prunus spp.

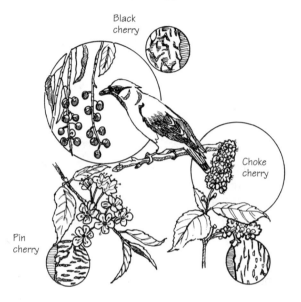

Black cherry

Choke cherry

Pin cherry

Horizontal lenticels, breathing pores that look like white slits on the bark, are characteristic of cherries (but are also common on birches and alders). The leaves are alternate, toothed, and single, the black cherry being most easy to identify by the rusty fuzz covering part of the midvein on the underside.

Cherry trees produce thousands of fruits. Humans find the raw fruits tart and not particularly edible, but just about anything else that moves prizes them. Robins, crossbills, waxwings, grosbeaks, starlings, flickers, grouse, brown thrashers, woodpeckers, black bears, fox, coyote, squirrels, deer, moose, and rabbits, all eat cherries whenever they can find them. The pits' stony indigestibility ensures that they are regurgitated or excreted somewhere distant from the tree where they may sprout and form a new stand of saplings. Most cherries also vegetatively reproduce by sending out root suckers, which turn up a short distance from the host tree.

Black cherries are numerous in young forests but survive poorly once the forest matures. The young trees can withstand shade, but the older trees require sun and die without it.

The Ojibwa gathered chokecherries, dried them on reed frames, and pounded them, stone and all. They ate black cherries and pin cherries raw, or cooked them and spread them on birchbark to dry. In winter, they boiled the dried berries and seasoned them with maple sugar, or combined them with other foods. The twigs of chokecherry and black cherry, tied in a bundle and dropped into hot water, made a fine beverage.

The wood of the black cherry is prized for its mellow red-brown tones and its durability. Black cherry trees can grow to be over one hundred feet tall and 3 feet in diameter, but a mature tree is a rarity—the wood is just too valuable, for everything from gun stocks to musical instruments.

OBSERVATION

During and after the August fruiting season, look for broken branches and long claw marks on the trunks, evidence that black bears have gorged themselves on the cherries.

Ironwood

Ostraya virginiana

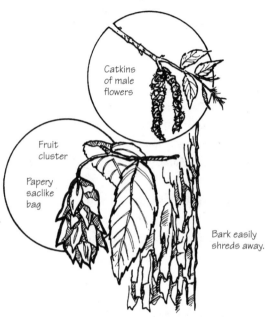

Catkins of male flowers

Fruit cluster

Papery saclike bag

Bark easily shreds away.

The double-toothed, simple, alternate leaves look similar to a number of other species, such as yellow birch, and so are difficult to use alone as a final identifying characteristic. But the bark helps in identification; its narrow vertical stripping shreds away from the trunk rather unlike that of any other understory tree. And the fruits give away the identity completely— the flattened nuts are each enclosed in a 1/2-inch long papery bag. The bags form an elongated cluster of sacks, each overlapping the other. The male and female flowers hang in separate clusters of cylinder-shaped catkins and appear along with the leaves in May. Ironwoods seldom grow over 30 feet tall or spread more than 1 foot in diameter.

Ironwood grows in the shade of larger northern hardwood trees like sugar maple, yellow birch, and basswood, and remains there as an understory member of older forests. Ironwood tolerates shade in much the same way as balsam fir, helping to form an intermediate layer of foliage between the tall canopy trees and the shrubs. As a general rule, the more horizontal layers in a forest stand, the greater the numbers and variety of wildlife. Even-aged, same-height forests, like red pine plantations, simply don't supply the rich habitats that forests of many ages, many heights, and many species can provide.

Ironwood lives in the richer deciduous soils of the North Woods, never appearing in dense stands, growing at a medium pace to a medium height over a medium life span, and offering no significant commercial importance. It's easily overlooked and under-appreciated, though the leaves turn a buttery gold in autumn, softening the middle heights of the taller forest. The wood, however, is extremely tough. At 51 pounds per cubic foot (compare to hemlock at 20 pounds per cubic foot), ironwood is heavy, extremely hard, and considered 30 percent stronger than white oak. In pioneer times, ironwood was the choice for tool handles, axles, and spokes, and as a firewood, it knew no competitors.

Many people refer to ironwood as hop hornbeam, the "hop" from the resemblance of the fruiting mass to hops, the "hornbeam" in apparent reference to the use of a similar European tree for oxen yokes. It is also called leverwood because of its remarkable strength as a levering tool.

The ironwood sheds its small seeds in winter, and

they blow along the surface of hard-packed snow—another adaptation helpful to the distribution of its seeds.

Ironwood has limited wildlife value—deer and rabbits browse it, and various songbirds, grouse, and squirrels eat the seeds, buds, and catkins. The Ojibwa used the young saplings for wigwam poles because of their strength.

OBSERVATION

When in any forest, note the number of layers of vegetation. Look for a healthy ground layer, low shrubs, tall shrubs, small trees, and large trees. Various birds have evolved to use the different levels of a forest for nesting, cover, and feeding; so, as a general rule, the more layers, the richer the bird life.

Black Ash

Fraxinus nigra

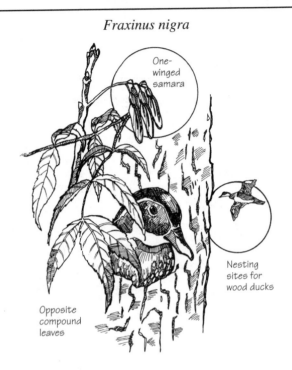

One-winged samara

Nesting sites for wood ducks

Opposite compound leaves

A shes are easily identified because they are just about the only tree in the North Woods with opposite compound leaves that are pinnate—the leaflets grow out from the center midrib (the long stem around which the compound leaves are all attached) in a featherlike pattern. Among tree species in the North Woods only maples, ash, and dogwoods have opposite leaves—the acronym MAD (maple, ash, dogwood) helps one to remember this triad. Green ash (*Fraxinus pennsylvanica*) occurs throughout the North Woods, and white ash (*Fraxinus americana*) appears in northern Michigan and eastward but seldom in northern Wisconsin and

Minnesota. Black ash leaves look quite similar to the other ashes, but the seven to eleven toothed leaflets have no stem (petiole) attaching them to the midrib, while the others have small stems joining them to the midrib. Black ash grows in wetlands and along banks of streams and lakes, in a shape maybe best described as bedraggled. Its trunk may be thick, but its overall shape is very slender, with a few short, coarse branches projecting irregularly upward to form a narrow crown. This differs greatly from the full crowns of the white and green ash, which are often planted for their shady canopies.

Black ash is the only important hardwood species characteristic of northern lowlands (other hardwoods may grow in lowlands, but grow better, in uplands). Found often in floodplains and swamps, black ash can tolerate standing water in part because of the new growth's numerous air pores, which can transport oxygen down to the roots. The shallow root system, typical of wetland species, makes it subject to being blown down in a good storm, but it often sprouts from the broken branches or stumps, and thus perseveres.

The flowers bloom in May before the leaves appear; black ash is the last tree in the North Woods to leaf out, waiting usually until June. After the first frost of autumn, which usually occurs in late August, the leaves quickly turn brown and fall, so for eight to nine months out of the year black ash displays bare limbs.

The fruit is a one-winged samara, blunt at both ends (other ashes' are pointed), ripening in September. The black ash seed looks like a single-winged helicopter, or like one-half of the maple samara but it is shaped like the blade of a canoe paddle.

Germination occurs in the second spring after fruiting, a rather unusual waiting period, and the seeds germinate best on moist organic soils.

The "black" in the name originated because of the blue-black winter buds and the dark brown heartwood, which often runs throughout the tree. The dark wood is heavy, tough, and durable, and is used for interior trim, cabinets, furniture, and veneer. Black ash grows huge burls that are prized for veneers and furniture.

Black ash also is known as hoop ash and basket ash, because if the ends are hammered repeatedly, the wood will split along the annual growth rings into thin sheets that can be cut into strips for barrel hoops, chair seats, and woven baskets. Native Americans made fish baskets from the strips.

Black ash offers moderate wildlife value. Purple finches, pine and evening grosbeaks, wood ducks, beaver, porcupines, and squirrels eat the winged seeds, and white-tailed deer and moose eat the twigs and foliage. Frequent browsing by deer may create a heavily branched bushy growth quite unlike the normal spindly form.

OBSERVATION
During the winter, look for ash seeds skittering across hard-packed snow, another of nature's ways of broadcasting seeds away from the parent tree.

Conifers

Eastern Hemlock

Balsam Fir

White Pine

Red Pine

Jack Pine

Black Spruce

White Spruce

White Cedar

Tamarack

Eastern Hemlock

Tsuga canadensis

Black-throated green warblers like nesting in hemlocks.

"Hemlock has a stemlock."

The hemlock's needles spread out in a flat spray like those of balsam fir; however, hemlock needles have a very short stem that attaches them to the twig, and balsam has no stem. Remember the phrase, "hemlock has a stemlock." Hemlock has a more lacy look than balsam.

Hemlocks are usually found in old-growth forests. Their ability to withstand shade allows 1-inch saplings to remain alive for one hundred years or more while awaiting their chance for an opening in the canopy. Some 2-inch and 3-inch saplings have been found to be two hundred years old.

Mature hemlocks can reach a record 160 feet tall and be 6 feet in diameter, but they average 80 feet tall and 2 to 3 feet in diameter. They are slow-growing and long-lived—the record hemlock reached 988 years, though 300 years would be more common.

Hemlock needles "podzolize" the soil, meaning they leach out the nutrients and ultimately create an acid soil that few plants can tolerate. When that feature combines with the deep shade cast by their dense canopy, a hemlock glade can often be devoid of any understory vegetation, and thus makes a parklike environment for walking.

Early lumberjacks cut hemlock not for the wood, but for the bark, which contains nearly 10 percent tannic acid. The bark-peeling season began in May and ran until mid-August. Cattlemen in Texas shipped their hides by boat to tanneries in northern Wisconsin towns like Mellen, Lugerville, Medford, Phillips, Prentice, Rib Lake, and Tomahawk. Most of these tanneries remained in business until around 1925, when other chemicals for tanning began to be used. In the early days, the hemlock wood was often left to rot on the forest floor, but it was later used for pulp and a poor-grade saw timber.

Deer are wild about hemlocks, and browse the seedlings voraciously. Without control of deer populations, hemlocks will continue to diminish in number, a significant aesthetic loss to humans and a greater loss yet to the veery, the black-throated blue warbler, black-throated green warbler, junco, and Blackburnian warbler, all of whom find hemlock a favorite nesting place.

ACTIVITY
1. Collect a handful of needles and steep them in boiling water for several minutes. The resulting tea is pleasant, though different from teas we are

accustomed to. Yes, Socrates did indeed die from drinking hemlock tea. But that European hemlock (Conium maculatum), though it's naturalized in the United States, grows as a 5-foot-tall bush—a far cry from the North Woods' majestic hemlocks.

2. On a sunny day, carry a light meter into a hemlock stand and measure the amount of light at ground level compared to the light at ground level in an open area. If you wonder why the hemlock stand has so few ground layer plants, analyze what the light meter tells you—most plant life needs sun, and hemlocks close out the sky.

Balsam Fir

Abies balsamea

2–4" cones
stand erect
on branches.

The flat, blunt needles, 3/4 to 1 1/4 inches long, are arranged in flat sprays, with no stem attachment to the twig. Balsam needles are coarser than the lacy needles of the hemlock.

Very shade tolerant and able to "wait its turn" for a gap in the canopy, balsam is a common understory tree in maturing forests. Balsams are often found in dense stands where their growth potential has been released by a cutting of the overstory trees.

The 2- to 4-inch cones of balsam fir stand erect on the branches, unlike those of any other conifer in the North Woods. When the cones ripen in the fall, the seeds fall away, often leaving the bare axis projecting upright on the branch.

Balsams grow in a perfect pyramidal shape with a

church spire tip, an ideal form for shedding heavy snow loads. Heavy snows are actually beneficial to balsams. When the lowermost branches are weighed down, they acquire a permanent droop. Wherever the branch contacts the ground, roots can develop and the branch may turn upward and become a new tree. A balsam can thus clone itself in successive circles.

In the boreal forests of Canada and the boreal stands that reach into northern Wisconsin, balsam is the dominant tree species, along with white spruce. However, balsam seems to prefer the moist sites while white spruce enjoys the dry sites. Balsam doesn't seem to mind germinating in heavy leaf litter on the forest floor, and is able to put down its new roots through 3 inches of decaying leaves.

Deer, moose, and snowshoe hares eat the twigs, but for deer, balsam is often a last resort. Porcupines eat the bark, but porcupines eat the bark of virtually any tree they bother to climb.

ACTIVITY

The smell of balsam fir needles is the essence of Christmas. An even more powerful distillation comes from the resin trapped in blisters on the surface of the bark. Pop one of the blisters with your fingernail to release the clear, very sticky liquid, a substance still used as a transparent cement for glass in optical instruments and microscope slides. The resin was once used for making torches and for healing cuts.

White Pine

Pinus strobus

Logging white pine in late 1800s

Mature white pine provide good nesting sites for eagles.

Five needles in a cluster (five letters in "white"—five needles in a bundle) is all you need to know. White pine also has a feathery, lacy look.

White pines speak an ancient language, one that can still be heard but only from the tiny remnant stands that chance or private ownership preserved. The sound of wind in an old white pine is like waves falling back along a sand beach, like a hare racing over powder snow, like a mother comforting a waking baby, "ssshhh, ssshhh." And the silence of a heavy snow in an ancient stand of white pines is like the lowest string on a viola, barely touched, thrumming low, low.

Although some relatively pure pine lands existed historically, white pines were never very numerous. An acre of North Woods old growth might average 150 trees, of which 12 to 15 white pines might have been scattered throughout. Although poets and loggers dreamed of a vast white pine forest, stretching from Maine to Minnesota, original surveyor reports, usually made 10 to 15 years after removal of the Native American tribes and well before the lumberjacks' arrival, commonly describe areas of open land with young growth. Fur traders' journals, many years prior to the surveyors, tell the same story. One traveler, on an 1847 journey from La Pointe in the Apostle Islands to Prairie du Chien, wrote that the flat or rolling uplands were thinly timbered, with the deep forests relegated to the valleys: "The hills are covered with a growth of small timber, mostly pine, with some sugar maple, oak, and a few aspens."

Nevertheless billions of board feet of pine were available to the saw, and the trees were culled with precision, for the pines could float down the rivers to market, but the hardwoods would sink to the bottom. The immense harvest still staggers the imagination.

The lumbering unfortunately wasn't the end of the problems for white pine. In 1909 a fungus, white pine blister rust, was introduced from Europe. It still today takes a large toll on young white pines.

And what about the few stands of big pines remaining? Their great height, also is often their downfall. The pines stand like large feathers overarching the forests, open to the high winds, and terribly vulnerable to windthrow.

Some of the white pines reported by the first surveyors dated back to the 1400s, and reached heights of 200 feet and diameters of 7 to 10 feet. They are trees of legend now, but one can only hope we will soon enact an ancient-tree protection bill that will allow distant generations the chance of seeing legends come true again.

ACTIVITY

Take a break and lie under a big old white pine. Feel the soft needle bed under you, and see the sun flecks that steal between the needles of the canopy. Listen for the sound of the wind, translated into pine music. Breathe deep. Don't be surprised if you fall asleep.

Since white pines usually are the largest of all trees in the North Woods, take along a measuring tape and a ruler to measure the granddaddy white pine in your area. Circle the tape around the tree at about your chest height (4.5 feet). The current Wisconsin record circumference for a white pine is 210 inches.

To determine the tree's height, stand far enough back that you can see both the top and the base of the tree between the 0 and 12-inch marks on a ruler held out straight from your body at eye level. Then move the ruler toward you until the tree fits exactly between the 0 and 10-inch marks. Keeping the ruler just as it is, now sight across the 1-inch line and have a friend mark the corresponding point on the tree. Measure how high this point is from the base of the tree, multiply by 10, and you have a good estimate of the height. If you find what you believe could be a record, contact your state DNR Bureau of Forestry.

Red Pine

Pinus resinosa

Red pine

Jack pine

Red squirrel eating on midden pile.

R ed pine has two needles in a bundle, and each needle is 4 to 6 inches long, creating a coarse look compared to the lacy needles of the white pine. The bark is a scaly reddish brown.

Red pines do quite well throughout the natural forest in the North Woods, outcompeting many other trees on sandy, gravely soil. Commonly 60 to 90 feet tall, they set records of 150 feet in height, 4 feet in diameter, and 300 years in age. Red pines flourish in open sunny conditions, and seed in after fires, often producing even-aged stands.

Red pines seem to be a forester's dream tree: a tree that grows straight and fast, likes full sun, self-prunes its lower branches, resists insects and disease, and is happy in poor sandy soils. Once harvested, the wood may be turned into framing lumber, flooring, poles, and pulpwood. All in all, red pine is a profitable species that requires a minimum amount of labor and worry.

But as with most things that look so good on the surface, red pine carries a price. The needles are highly acidic and decompose slowly, gradually degrading the soil in which they grow. During the current century this podzolization has cut the productivity of European conifer plantations in half. Red pine plantations are often "ecological deserts," devoid of shrub or ground layer species. While red squirrels, which eat the seeds of red pine, and porcupines, which eat the cambium, thrive in a red pine plantation, most other wildlife species can't find a meal there.

Forest plantations disregard a simple ecological maxim, that nature abhors sameness. Planting one species over large tracts of land turns the forest into a farm and reduces the overall long-term health of the system. Nature's stability lies in diversity of organisms. The more organisms at all levels of the forest, the less chance of a total system collapse. City foresters learned this lesson when they planted elms to the exclusion of most other species, and then watched them die when Dutch elm disease spread down the line of trees on block after city block.

Aldo Leopold wrote that pine needles "are filed in the duff to enrich the wisdom of the stand. It is the accumulated wisdom that hushes the footsteps of whoever walks under pines." That wisdom also tell us that when forests with red pines also include a diverse array of other trees and shrubs, those habitats remain healthier over the long term.

ACTIVITY

Grab a handful of soil in a red pine plantation, and feel how sandy it is. Soil structure determines how fast water will percolate in and through a soil. Sand drains rainwater through in a hurry, creating a dry habitat. Water doesn't soak into clay, it runs off over the top. Loams absorb and hold onto rainwater, releasing it slowly. To determine how fast water percolates through a soil, cut the bottom and top out of a tin can, twist it firmly into the soil, fill it to the top with water, and time how long it takes to empty out. Compare the results of this test in a series of different soil types.

Jack Pine

Pinus banksiana

Jack pines have adapted to the least fertile soils.

Cones open only when 116 degrees is reached.

Like red pine, jack pine has two needles in a bundle, but they are usually only 1-1/2 to 2 inches long, much shorter than the needles in red pine. Incurved cones also distinguish jack pine from white and red pine, both of which have straight cones.

The most northern of our pines, reaching far into Canada, jack pines love full sun and the open habitat that follow forest fires. Their cones are sealed tight by resin that softens and allows the cone to open only when 116 degrees is reached. The cones do not drop off the tree every year like the cones of other pines, but remain up to twenty-five years awaiting a fire or a day of scorching summer heat. One study reported up to two million seeds per acre waiting in reserve in unopened cones. The seeds usually can withstand the heat of a fire long enough for the fire to push by, and the bed of ashes left behind offers a perfect germinating site. The young saplings grow rapidly and within five years produce seed themselves, in anticipation of the next fire.

Jack pines are the poor cousin of the red and white pine. They are adapted to the least fertile soils the North has to offer (which are as poor as one can find—very nearly pure sand). They live a short life and are often scrubby in comparison to the majesty of the big red and white pines (though they can reach 80 to 100 feet and be straight as an arrow). Remarkably, jack pines can be found in the middle of bogs, and they can also colonize rock outcroppings with a bare minimum of soil and moisture.

Jack pine stands provide the sole habitat supporting Kirtland warblers, an endangered species that requires fifteen- to twenty-year-old jack pines for nesting. Deer, snowshoe hares, and porcupines readily browse jack pine foliage.

ACTIVITY

Collect a handful of sealed jack pine cones, and place them on a piece of white paper in a microwave for thirty seconds. Watch as the cones unfurl like a blossom. Shake the cones and collect the loose seeds for planting in the spring.

Black Spruce

Picea mariana

Black spruce is a dominant tree in bogs.

Black spruce has shorter needles, 1/4 to 1/2 inch long, than white spruce and smaller, rounder cones. All spruce needles roll in your fingers even though they are four-sided; balsam fir and hemlock are flat and won't roll. Spruce branches look like a bottle brush because of the way the needles wind tightly all around the woody stems.

Black spruce generally wins no prizes for grace of design, majesty of size, or beauty of foliage. In fact, often they look almost dead. Narrowly pyramidal in shape, and often with all the lower branches dead and covered with lichens, many black spruce show tufts of live growth only at their tops. But on dry upland sites, black spruce can be handsome, commonly reaching 60 feet in height and a foot in diameter.

In the bogs, black spruce and tamarack are the dominant trees, though sometimes jack pine and red maple somehow also survive in the sphagnum moss. It's not unusual to find a 6-foot-tall spruce that is 1 inch in diameter and seventy-five years old. The cold, acid sphagnum moss makes conditions virtually impossible for most plants to survive. But black spruce manages, and has developed a few adaptations to help it along. Seed cones are produced annually, but may hold on to their seeds and their spot at the top of the tree for thirty years until a fire pokes its way through the muskeg and triggers the seeds to disperse. Huge stands of even-aged black spruce attest to this patient wait for fire.

Layering, a process in which the lower branches hang close to the ground and slowly become covered with organic matter, is another key method of black spruce reproduction. At the point where a branch contacts the soil, it sends down roots. Then the branch turns upward, and it eventually becomes the trunk of a new tree, a handy trick if seeds have trouble surviving in the sphagnum.

The Ojibwa and the French voyageurs used black spruce roots as the lacing for their birchbark canoes.

OBSERVATION

Look for "fairy circles" of black spruce, a circle of smaller saplings around a larger single tree, produced by layering of the larger tree's lowest branches.

White Spruce

Picea glauca

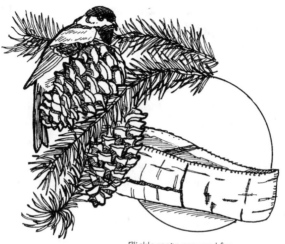

Pliable roots are used for sewing up canoes and spruce gum for caulking.

White spruce needles are 1/2 to 3/4 inch long, a bit longer than black spruce's. The needles are four-sided and rollable between your fingers like black spruce's, but white spruce is seldom found in bogs like black spruce.

White spruce is truly a tree of the North, ranging far up into Canada and Alaska and acting as the dominant tree species, along with balsam fir, of the boreal forest. It can handle the minus-60-degree temperatures and the heavy snow loads with relative ease. The tightly wound waxy needles hold moisture in, preventing desiccation and death in winter winds.

Spruce use extracellular freezing to survive the cold. To prevent damage from freezing and bursting, spruce cells ooze liquid out through their walls into empty spaces between the cells. Here the liquid can freeze without harming the living cells. Without this adaptation, trees adapted to the subarctic, like white and black spruce, jack pine, balsam fir, tamarack, paper birch, and aspens, would literally burst when the temperature dropped to minus-40 and below.

White spruce usually bears a good cone crop every three to five years, and after a summer burn, seeds will sprout on the newly exposed soil. Some years a white spruce will produce few cones or none at all. Two healthy spruce standing side by side may reflect this cone feast or famine—one laden with cones and the other bearing none.

When crushed, needles of the white spruce don't smell very good, so this variety is not a favored Christmas tree. A host of seed-eating birds, red squirrels, chipmunks, and mice like the seeds.

Traditional Native American uses included sewing up canoes and baskets with the pliable roots, using the twigs as a medicine for stiff joints, and applying spruce gum (considered the best pitch of all the conifers) to caulk canoes and birchbark pails.

Activity

Collect a bit of dried "gum" (pitch) from the exterior of a white spruce and chew it like gum. It sticks to the teeth and tastes pretty strong, but it was historically used for gum.

White Cedar

Thuja occidentalis

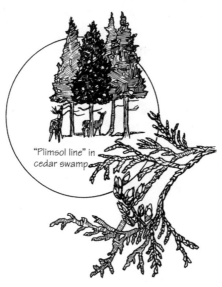

"Plimsol line" in cedar swamp

The flattened needles, with scales that overlap like shingles on a roof, easily distinguish cedar from other North Woods trees. Cedar bark shreds in vertical strips, an additional identifying characteristic.

Arborvitae, Latin for "tree of life," is another name for white cedar, the name originating from the tree's use by Jacques Cartier in the 1500s as a remedy for scurvy. But an equally good reason for the name is cedar's ability to continue growing after it has been knocked over by wind or snow. Even though the tree may be lying flat on the ground, new roots will grow down from the horizontal trunk, and also from branches now shooting straight up into the air. Straight rows of young trees are common in cedar swamps because of this adaptation.

White cedar also uses layering. Lower branches hang so low that they often get covered by mosses and organic debris. They then send out roots at their tips, the branch turns upward, and a new trunk begins. "Fairy rings" of cedars are not unusual, because of this reproductive strategy.

Cedars also produce seeds, encapsulated in a rather tiny cone, about 1/2 inch long. The seeds seem to germinate best on rotten logs and stumps, which when they rot fully away leave the cedars "standing on their toes," ready to fall in a good wind.

White cedar grows in wet, spongy soils, usually in tangles of roots, trunks, and fallen trees that make walking a bit tricky. But white cedar also thrives on rocky sites, particularly limestone. Black River Falls near Lake Superior is a rock cliff corridor with white cedars balancing on precipitous ledges, clinging to rock where there is no apparent soil for rooting.

OBSERVATION

White-tailed deer love to gather in cedar swamps for the winter. The dense growth cuts off the wind almost entirely; the snow gets caught in the upper branches, reducing heavy accumulations on the ground; and the foliage is a highly desired browse. "Plimsol lines," almost perfect lines of foliage at the top height a deer can reach, are common, highly visual results of excessive deer browse.

Tamarack

Larix laricina

Tamaracks are dominant bog dwellers.

Tamaracks really can't be confused with any other tree in the North Woods. The twelve to twenty needles in a bundle growing in a starburst on a raised spur makes tamarack unique. Tamarack is our only deciduous conifer, turning gold in the autumn and losing its needles (though all conifers lose their needles over the course of several years).

Tamaracks are dominant bog dwellers along with black spruce, somehow having developed a liking for the cold, acid, nutrient-poor conditions a bog offers. In the sphagnum moss mat of the bog, roots penetrate only a minimum depth. Tamarack responds by producing new roots from the main trunk, above earlier roots. Tamarack also vegetatively reproduces (clones itself) by sending out root sprouts, which can become new trees. If all else fails, every four to six years it produces seeds in massive quantities, numbering up to five million per acre (about one hundred per square foot) in a mature stand.

Tamaracks need full sun to survive, and won't grow in their own shade. A bog, with its very poor growth conditions, is a good place to find lots of sun. Those trees that do survive usually grow very slowly. Tamarack will also grow on more upland sites, and with the better conditions, can fill out more quickly.

Young tamarack, because of its strength, was one of the Ojibwa's two favored choices for wigwam poles (ironwood was the other). They used the inner bark as a medicine for burns and the roots for weaving bags and sewing canoe edges.

The wood is prized for its toughness and resistance to moisture and weather. The bark contains tannins useful in tanning leather. Porcupines like the inner bark, often girdling the tree and killing it.

OBSERVATION
Tamarack needle clusters emerge from peglike woody knobs all along the branches. In winter, when the needles have dropped, the mosquitoes are quiet, and the bogs are iced over, take a look at this unique identifying characteristic.

Shrubs

Staghorn Sumac

Blackberries/Red Raspberries

Red Osier Dogwood

Beaked Hazelnut

Highbush Cranberry

Willow

Tag or Speckled Alder

Blueberry

Sweet Gale

Maple-leaved Viburnum

Thimbleberry

Sweet Fern

Staghorn Sumac

Rhus typhina

Sumac colony
in winter

Staghorn sumac reaches up to 15 feet tall, with alternate, long, compound leaves and eleven to nineteen toothed leaflets. The fruits in the fall are very large (up to 1 foot long), bright red clusters of fuzzy seeds. The twigs and leaf stalks look and feel velvety, much like the downy antlers of a buck "in velvet," hence the name staghorn. Note: Poison sumac is found in southern counties of Wisconsin, Michigan, and Minnesota, and generally need not be a concern in the North Woods.

Staghorn sumac blazes red, orange, and purple along many highways in the autumn, and because it grows in colonies, the visual display can be beautiful. The colonies result from sumac vegetatively reproducing through root suckers. Sumac needs full sun and doesn't mind poor, dry soils, so it's common along roads, in fields, and on hillsides.

Sumac had a host of historical uses. The cured leaves were a common tobacco mix for American Indians, the large fruits made a cool drink with a sour zing to it like lemonade, and black ink is reported to have been made from the boiled leaves and fruit. The tannin found in the leaves and twigs was used in tanning fine leathers. The Ojibwa used the flowers for a stomach pain remedy and the pulp of the stalk and inner bark as a dye. The Menominee made an infusion of the inner bark for hemorrhoids. The Micmacs drank a tea made from the berries for sore throats. The Latin name *typhina* originated in the belief that sumac had value as a cure for typhoid fever. And young stems were cut, the pith removed, and the resulting hollow spiles used for collecting maple sap.

Sumac provides a particularly good winter source of food for wildlife because the fruits hang on well past the new year. The fruits are a good source of provitamin A and have been found in the stomachs of birds like ruffed and sharp-tailed grouse, mourning doves, crows, and a host of North Woods songsters such as bluebirds, flickers, catbirds, phoebes, robins, and thrushes. The twigs are browsed by deer, cottontail rabbits, snowshoe hares, and moose.

ACTIVITY

Make sumac lemonade by steeping the fruit in boiling water and then cooling. Or make sun lemonade by simply leaving the fruit in a jar of water under the hot sun for an afternoon. Add honey to reduce the tartness.

Blackberries/Red Raspberries

Rubus spp.

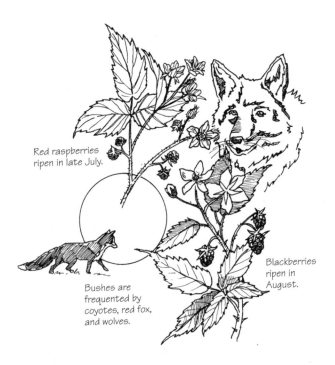

Red raspberries ripen in late July.

Bushes are frequented by coyotes, red fox, and wolves.

Blackberries ripen in August.

Blackberries and raspberries grow on tall (3 to 8 feet) arching shrubs with compound leaves, each with three to seven toothed leaflets, usually three in a feather shape on raspberries and five in a fan shape on blackberries. Both species have prickles; but raspberries' round, bristly stems, like stiff hairs, have a hard time drawing blood, whereas blackberries' angular, stoutly barbed stems do an excellent job of ripping skin and clothing.

Both produce white flowers in May and June. In a good year, raspberries provide a wealth of fruit in late

July, and blackberries offer their bounty in August. Every year they produce new canes, each living a year or two and bearing fruit in the second year, while the roots live on year after year.

No species name is given for these plants (represented by "spp.") because as one early botanist put it, "This group is especially to be commended to the systematic botanist who is seeking problems." Both species hybridize often enough that identification is difficult.

Raspberries and blackberries are stoloniferous, meaning they send out aboveground stems from the parent plant that develop into new shoots and quickly into thickets. And both are sun lovers, willing to tolerate poor soils. As any berry picker knows, you look along old roads, in old fields, around recent burns, and in new logging cuts for the best picking and largest berries.

About 150 species of birds and mammals, representing nearly every species on the continent, share the human appetite for raspberries and blackberries. The thorny brambles also make an effective cover and nesting site for birds, and small mammals often hide in such thickets, as well as eat the stems as a browse in the winter. Wild dog family members (coyote, fox, and wolf) eat berries too. In fact, about one-fourth of the summer and fall diet of a red fox consists of fleshy fruits and seeds like blackberries, wild cherries, and blueberries.

Humans enjoy raspberries and blackberries, fresh with cream and sugar, sprinkled on cereal or ice cream, and baked into a variety of pies and other desserts. The leaves can be made into mild teas. The Ojibwa used them along with the roots as medicines for lung troubles, eye diseases, menstrual discomfort, and dysentery.

ACTIVITY

Why make life complicated? Simply pick, eat, and enjoy. Watch for the flowers in bloom to determine where the picking will be best later in the summer.

Red Osier Dogwood

Cornus stolinifera

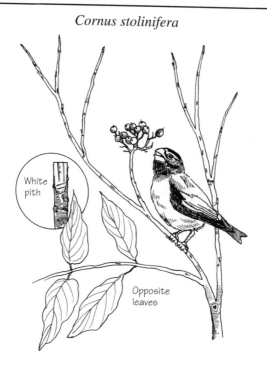

White pith

Opposite leaves

The shiny, fire-engine red bark on the twigs and branches distinguishes this dogwood from all other shrubs in the North Woods (though the stems may be more greenish in summer). As with all dogwoods, the leaves are smooth along the entire edge, the veins all curve up toward the tip of the leaf, and the simple leaves are opposite (all dogwood leaves are opposite except, obviously, the alternate-leafed dogwood). Cut through a twig, and the pith is white. Red osier dogwood flowers are small, white, and not particularly fragrant. Arranged in flat-topped clusters, they appear in late May to June. The white berries develop in July and last into

October if the birds don't get them earlier. This 4- to 10-foot tall shrub seems to enjoy wet feet, with its shallow roots well adapted to nearly all wetland habitats, like shores of rivers and lakes, marshes, and sand dunes.

The scarlet red branches of red osier dogwood supply a welcome splash of color in a winter wetland usually clothed in sheer white. The lower branches tend to droop and become covered with organic material. The branches then take root and form a clonal shoot of the parent. The Latin species name *stolonifera* means "bearing stolons," in reference to the elongated horizontal stems that grow along the ground and take root. Large colonies may appear in a wetland, all relatives of one another through their vegetative reproduction.

Sexual reproduction works for red osier dogwood too. The striking white berries have a high nutritional value; a host of bird like thrushes, grosbeaks, grouse, and robins seek them out and then spread the seeds. Dogwoods rate fifth in wildlife value among North American plants. The fat content of their berries pulls in flocks of migrating birds needing a light but full calorie tank for their flights. The fruit isn't the only draw; the deer, moose, rabbits, and hares browse the twigs.

"Osier" is a term for tough flexible branches, and usually indicates a species well suited for basketry.

Beaked Hazelnut

Corylus cornuta

Red female flower

The most common shrub of northern dry forests, hazelnuts grow in a clump of slender stems 4 to 12 feet tall, with alternate, finely toothed, heart-shaped leaves that turn bright yellow in the autumn. The beige male catkins emerge in the fall and hang all winter until they expand in the spring to release their pollen. Hazelnut blossoms in very early spring, as early as any plant in the North Woods, the female flowers emerging from the buds in beautiful tiny scarlet hairs (the stigmas). The fruits identify the shrub—bristly husks with long tubular beaks enclose edible nuts that often grow in twos like a pair of wings. The bristles on the fruit husks are tough enough to poke into the skin and break off.

Thickets of hazelnut develop in woods because hazelnut sends out underground rhizomes that clone into new shrubs, a form of vegetative reproduction particularly necessary for hazelnut because few of its nuts get past the squirrels and chipmunks. In older woods, hazelnut clusters become large stands. The dense growth not only offers food, but provides good cover and nesting sites for many mammals.

Squirrels, chipmunks, jays, deer, grouse, and mice vie for the rights to the nuts; seldom are any left long enough to be picked by humans. The nut has exceptional nutritional value, with 25 percent protein and 60 percent fat. Grouse eat the catkins, and rabbits, deer, beaver, and moose browse the whole plant, though hazelnut is not considered a particularly good winter browse.

People can eat the nuts raw, but they're better when roasted, or they can be ground into flour after roasting. The filbert of commercial trade closely resembles the hazelnut and belongs to the same family.

OBSERVATION
In April, when the snow still covers the ground, check the buds on hazelnuts for the tiny scarlet female flowers, a true harbinger of spring.

Highbush Cranberry

Viburnum trilobum

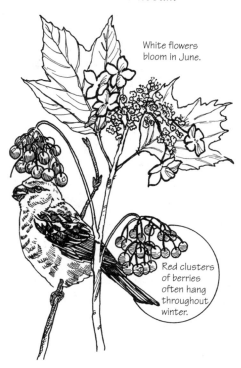

White flowers bloom in June.

Red clusters of berries often hang throughout winter.

An 8- to 17-foot-high arching shrub that grows mostly in and along wetlands, the highbush cranberry has distinctive opposite, toothed leaves. The three long-pointed lobes (trilobum) give the appearance of a red maple. The fall leaves turn a lustrous scarlet.

In June, the white flowers bloom in umbrella-shaped clusters measuring 3 to 4 inches across. The outer, larger flowers are sterile (they have no pistils or stamens), while the inner, smaller flowers are fertile. Perhaps the outer flowers serve as the billboard advertisement to wandering insects seeking pollen. The soft fruits begin as a yellow berry in September, eventually turn a brilliant red, and hang in drooping translucent clusters well into the winter.

The shrubs often keep their fruits all winter with nary a soul eating them. The books say bears, foxes, squirrels, chipmunks, grouse, thrashers, thrushes, starlings, grosbeaks, and cedar waxwings favor highbush fruits, but nothing touches the shrubs in the wetlands below our home. The twigs and leaves are seldom browsed either. Highbush cranberry belongs to the honeysuckle family and is unrelated to the familiar bog cranberry; the cranberry reference originates from the fruit's minor similarities to the bog cranberry. Some people make an exceptional jelly from the cooked fruit, but while the fresh berries are edible, they are rather bitter. The berries are said to be high in vitamin C and historically were used to prevent or cure scurvy.

The Malecites and Penobscots drank a tea from the highbush (which part is unknown) to cure mumps. Cranberry bark was known as "cramp bark" because of its use as a sedative and antispasmodic.

OBSERVATION

Note how long into the winter the fruits of highbush cranberry remain. Since highbush cranberry fruits have a low fat content, they rot very slowly, hanging around until late in winter when all the higher-quality foods have been exploited and they are all that's left. Then the winter birds and mammals turn to them and consume and disperse the fruits. Some plants produce high quality fruit that will rot quickly if not eaten. Other plants produce poor quality fruit that does not rot. Fortunately this fruit lasts until the toughest months of the winter when it is desperately needed by wildlife. So, both high quality and low quality fruits help support the winter survival of wildlife.

Willow

Salix spp.

Male flowers

Willow leaves generally grow long and thin, and are pointed at the tip, or lanceolate (like a lance). All willows have simple alternate leaves, and their buds are covered by a single scale that forms a hoodlike covering. The buds lie pressed against the twig, another unusual feature, and appear in midsummer, remaining over the winter until spring. With over fifty species of willow in the eastern United States (three hundred species in the genus), many of which hybridize or are highly variable, identifying specific species of willow tries the patience of the best of botanists.

Willow flowers produce nectar and colorful pollen. The early spring flowering comes and goes rapidly, and by late May or early June, the catkins have already gone to seed. Flask-shaped capsules enclose the seeds, and when the capsules break apart, they release numerous seeds winged with silky down that drift on the slightest wind, often creating a spring seedstorm. Willow seeds are tiny, remain viable for only a short time, and require moist sites for germination; thus they are usually restricted to germinating in and along stream banks and wetlands.

Willows grow in thickets, producing long underground shoots (stolons) from which new plants grow fast—up to 4 feet in a year. The fibrous roots spread widely, holding together the soil in wetlands and along shorelines. Many riverbanks owe their continued existence to the erosion control provided by willow roots.

Willows have a well-known reputation for sprouting—even stakes of green branches driven into the ground often will sprout—and shoreline reclamation projects for years have made use of the willow's survival ability.

The strong and flexible stems make superior baskets, brooms, and wickerwork. The greatest economic usage of willows is as a source of osiers (highly supple branches), but the bark has been used as a pain reliever since Dioscorides, a Greek physician in AD 60, prescribed it. Various Native American tribes used a tea from the wood to soothe arthritis and reduce fevers. The bark tastes bitter due to the presence of salicine, from which salicylic acid, the active ingredient in aspirin is derived.

Willow has high wildlife value. Grouse and pine

grosbeaks eat the winter buds, and snowshoe hares, beavers, white-tailed deer, muskrats, porcupines, and moose browse the twigs. The dense summer thickets provide nesting sites and cover for wildfowl and songbirds. A recent banding study of migrating songbirds near Park Falls, Wisconsin, showed that wetland thicket habitats are used as stopover sites by nearly sixty bird species—the top habitat of all those studied (In contrast, only six species use red pine plantations.)

ACTIVITY

Cut a stem of willow shrub and simply stick it into the ground in another wet site—it should root and bear leaves eventually. Or over winter, when it seems like spring will never come, cut a stem of willow and place it in a vase of water. The buds will open, leaves appearing from some buds and flower catkins from others, bringing spring into your home while the snow still flies.

Tag or Speckled Alder

Alnus rugosa

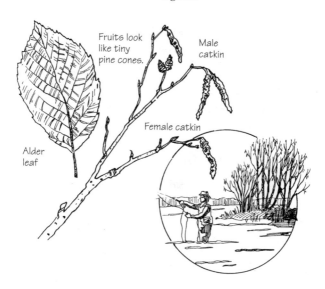

Fruits look like tiny pine cones.

Male catkin

Female catkin

Alder leaf

This tall (usually 8 to 20 feet high), wetland shrub has bronze-brown stems dotted with white linear lenticels or speckles. (Hence the name speckled alder). The leaves are finely toothed and elliptical, but the stalked buds provide a defining trait, as do the dark fruits, which look like tiny pinecones and remain on the plant nearly all year around. The male catkins form in the fall and wait out the winter until they expand with pollen in the early spring. The early spring flowers blossom before the leaves appear and offer little color, the males hanging in long catkins and the females projecting minutely in reddish hairs from short catkins.

Tag alders usually grow in clumps of six and seven stems, closely bordered by other clumps. Each cluster "locks elbows" with its neighbors.

Alders can clone themselves by sprouting new shoots from the base of an older trunk, sending root suckers underground and then shooting up new stems, or sprouting new shoots and roots when a branch leans over and contacts the ground (layering). All of this entanglement helps to hold soil in place along waterways where it might otherwise wash away, though it also aggravates anglers. Not only does alder control erosion, but, much like peats, beans, and alfalfa, it fixes nitrogen from the air and thus improves the soil.

Alders offer cool shade and dense cover for a host of birds and mammals, safe from human intrusion. The rich black muck soil provides worms and insects for the probing bills of woodcocks and snipes. The entangling branches supply nesting sites galore for red-winged blackbirds, goldfinches, and other thicket-nesting birds. The branches often lean out over a stream, offering shade and cover for trout, as well as a serious challenge for fly anglers.

Alder's wildlife value as a food is low overall, but its seeds are important in the diet of redpolls, siskins, chickadees, and goldfinches. Grouse eat the catkins and buds; muskrats, moose, and snowshoe hares feed on foliage and twigs; and beaver eat the outer cambium layer and then use the peeled sticks for their dams and lodges.

OBSERVATION

With a hand lens, look closely at the speckles all along the bark of alder. The speckles are lenticels, spongy air pores, which admit oxygen to the stems.

Blueberry

Vaccinium angustifolium

Flowers bloom mid-May into June.

This low, many-branched shrub has small, short-stalked, alternate leaves that are usually pointed and toothed, though some varieties are smooth edged. The slender twigs often zigzag back and forth, and are covered with "warts," little raised speckles best seen with a hand lens. The small flowers look like hanging lanterns and bloom white or faintly pink, usually from mid-May into June, while the fruits are often ready for picking in July. The berries, first green, slowly redden, and eventually turn blue-black, often with a whitish bloom. The seeds within the fruits are so tiny that most people don't even notice them.

Depending upon the species, blueberries grow in various habitats from old fields to forests to wetlands. I most often see blueberries in dry, sandy soil, usually in jack pine communities or sandy fields. But I also see them on the edges of bogs and on sandy beaches along lakeshores, wherever poor acidic soils and full sun occur.

Blueberries produce stolons, low stems that root at their nodes. Large colonies often result from this vegetative reproduction, all to the great pleasure of the multitudes of animals and people that await their ripening. Scarlet tanagers, bluebirds, robins, brown thrashers, and catbirds vie for the fruits, along with black bears, chipmunks, mice, foxes, and raccoons. White-tailed deer browse the twigs and leaves in winter, and cottontail rabbits eat both fruit and twigs. High in sugar but low in fat, blueberries supply adequate but not superior nutrition. The minute seeds usually pass right through most fruit eaters' digestive tracts and are thus distributed throughout the countryside.

OBSERVATION

Scout out the most likely spots for blueberries by looking in dry, sandy habitats like old fields and jack pine barrens, or along higher ground in a bog. Most people find that the wild berry, though small, far exceeds the flavor of the domestically cultivated varieties.

Sweet Gale

Myrica gale

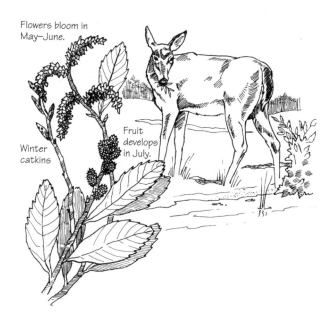

Flowers bloom in
May–June.

Fruit
develops
in July.

Winter
catkins

S weet gale is a 3- to 6-foot-high shrub with narrow
wedge-shaped alternate leaves that become toothed
near the rounded tip. The flowers bloom in May to June
in rather inconspicuous short catkins and develop into
equally inconspicuous small nutlets in July. The most
distinguishing characteristic of sweet gale is its spicy
fragrance, coming from yellow resin dots on the leaves
(use a hand lens to see them well). In winter, a season
usually bereft of many natural odors, if you rub sweet
gale's buds and catkins, the strong scent will burst forth.

Sweet gale grows along the edges of lakes, streams,
and wetlands, serving as an important buffer and stabi-
lizer between the water and land. By root suckering,

sweet gale vegetatively clones itself, sending up new shoots over large areas, and often extending out into the water. Not only does sweet gale stabilize shorelines, but its ability to fix nitrogen from the atmosphere through its root nodules improves the soil it is anchoring. Sweet gale is thought to have played an important postglacial role in beginning the process of building fertile soils.

The leaves were used historically as a meat flavoring, a clothing scent, an insecticide, a substitute for hops in northern Europe, a remedy for "the itch," and a vermifuge (worm expeller). Canadian Indians used the buds to dye porcupine quills. The leaves, when placed in drawers, were said to discourage moths. Deer browse the foliage and twigs, and sharptail grouse eat the buds and leaves.

ACTIVITY

Steep a handful of sweet gale leaves in boiling water for a pleasant tea, or chop up the leaves as a sage substitute for flavoring meats.

Maple-leaved Viburnum

Viburnum acerifolium

Somewhat flattened fruits ripen in September.

This small (3 to 6 feet), graceful shrub has opposite, toothed, three-lobed leaves that are rather velvety underneath and look much like the leaves of a high bush cranberry or a red maple. The white flowers blossom in late June in an upright, thin umbrella cluster, and ripen in September into first red, then purple-black, somewhat flat fruits.

Maple-leaved viburnum prospers in the rocky shade of dry deciduous forests, often growing in clusters due to its rootstocks' vegetative reproductive abilities. Its beauty stands out in late June when the creamy flower clusters are near eye level, in early autumn when the fruits turn deep purple and stand above the leaves in a candelabra bouquet, and a few weeks later when the leaves turn red-purple.

As handsome as the fruits are, their taste leaves much to be desired, as does their nutrient content. They often remain on the branches long into the winter, though grouse seem to enjoy them, while deer and rabbits browse the twigs. A New Jersey study showed 70 percent of the fruits still on the shrubs as of January 1, the fruits having ripened in August. The low lipid (fat) content (less than 10 percent by weight) makes maple-leaved viburnum relatively undesirable until later in the winter, when any food is at a premium.

Native Americans used the wood for arrow shafts because of its straight growth.

Thimbleberry

Rubus parviflorus

Thimbleberry grows along borders and trails, liking sunny areas.

Thimbleberry, a 3- to 6-foot-tall shrub, has very large (up to 8 inches wide), lobed maplelike leaves that are alternate (maples have opposite leaves). The only other shrubs with alternate maplelike leaves are currants and gooseberries, but their leaves are much smaller. The five-petaled white flowers appear on thornless branches and are often nearly an inch across, blooming in June. The red fruits appear in August, and when pulled off the stem are rounded like a bowl, or like a thimble—they do in fact fit nicely on the tip of a finger.

Thimbleberry grows in hardwood forests, especially at the edge of the woods, along trails, and in clearings where it can get some sun. Like other members of the bramble family, it often forms large thickets because its underground roots send up shoots that become new plants. Unlike many other members of the bramble family, though, thimbleberry has no thorns. In a colony of thimbleberries, the large leaves form a closed canopy over the ground, effectively shading out competition.

The species name *parviflorus* means "small-flowered," a completely inaccurate description since the flowers are really quite large. Thimbleberry also goes by the name salmonberry, in reference to the pinkish color of its fruit.

Eaten raw the fruit is mildly tart, dry, and modestly flavorful, a fine fruit to use in jams and tarts. Native Americans pressed the dried fruits into cakes and ate the young shoots as a vegetable. Over sixty species of birds, as well as bears, squirrels, porcupines, deer, rabbits, moose, hares, beavers, chipmunks, and skunks, eat the fruits from the genus *Rubus* (which includes raspberries and blackberries).

Activity
Pick a cup or two of thimbleberries in August and use them on your pancakes for a topping that will get some notice.

Sweet Fern

Comptonia peregrina

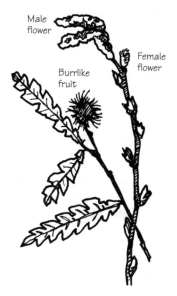

Male flower

Female flower

Burrlike fruit

This flowering shrub, usually 1 to 3 feet tall and heavily branched near the top, is not really a fern, despite its name. It has slender, round lobed, often sticky leaves that resemble a fern. Smell the highly aromatic leaves by crushing one in your hand; that sagelike scent easily identifies sweet fern and gives it its other common name, Indian sage. Many people, when the aroma is called to their attention, remark that they've smelled that scent for years but never known what it was.

Sweet fern produces separate male and female flowers in the spring before the leaves appear, the male flower clustered at the end of the branches in short catkins, and the female flower found below. The fruit is a green burrlike structure that looks a bit like a mace, a medieval weapon with a spiked metal head.

In the sandy habitat around my home in northern Wisconsin, sweet fern thrives, growing in poor soils where few plants can survive in. One can tell at a glance the quality of a soil by the presence of sweet fern. But sweet fern fills a necessary niche. By colonizing logging roads, old fields, and other disturbed sites, sweet fern helps to stabilize the soil. Sweet fern also fixes nitrogen, meaning it can convert nitrogen from the air rather than taking it from the soil, and so improves the soil in which it grows. Like other plant pioneers, sweet fern makes the bed for the growth of plant species that follow.

Sweet fern once had a host of medicinal uses: as an external liniment for bruises and for rheumatism; as an internal remedy for colic, diarrhea, and dysentery; and as a tea. The Mohegans applied the cooled tea to cure the rashes of poison ivy. Some people have used it for bronchial ailments, and its "Vicks"-like smell would seem to support that usage. The dried leaves were once used to scent pillows and clothing, and to flavor meat.

Ruffed and sharptail grouse, cottontail rabbits, and deer browse the buds and foliage. But sweet fern probably offers more value to wildlife as a shelter for small game than as a food, since its leaves are dense above but thin below.

ACTIVITY

To make the tea, simply boil some of the leaves— but not for too long, or the tea achieves industrial-grade status, making the eyes water. Or crush the leaves to use as a sage substitute.

Early Spring Wildflowers

Trailing Arbutus

Wood Anemone

Marsh Marigold

Blue-bead Lily

Hepatica

Gaywing

Spring Beauty

Bloodroot

Trout Lily

Sessile-leaved Bellwort

Trailing Arbutus

Epigaea repens

Trailing hairy stem

Stretching along the forest floor up to 15 feet in length, trailing arbutus is a tough woody vine with hairy branches and alternate, oval, thick leaves that are leathery and evergreen. The flowers blossom in pink and white waxy clusters, each flower 3/4 inch across and forming a tube with five petals flaring out. Few flowers are as intensely fragrant, and on occasions when arbutus blooms with the snow still on the ground, the sweet scent brings the message of spring.

In 1880 William Gibson claimed, "No other flower can breathe the perfume of the arbutus, that earthy, spicy fragrance, which seems as though distilled from the very leaf-mould at its root."

In New England trailing arbutus was called Plymouth Mayflower, and was sold on the streets by peddlers. It was the first flower seen by the Pilgrims after their first winter in the New World. Because it rises out of the dead leaves of the winter with such immaculate beauty and fragrance, the trailing arbutus came to symbolize the Pilgrims' ability to rise from peril and live again.

The Greek genus name *Epigaea* means "on the earth," referring to its trailing habit, while the species name *repens* means "creeping." Most commonly a resident of nutrient-poor pine forests, the arbutus puts down roots that grow in association with mycorrhizal fungi that help make nutrients available to the plant.

ACTIVITY

Keep track of the blossoming dates of the first spring flowers. Over the years the dates will vary, due to changes in soil, temperature, moisture, and light conditions, as well as other ecological factors.

"Phenology" is the study and recording of natural events from year to year and from place to place. It provides the answer for questions like "When do the blackberries ripen?" or "When are bear cubs born?" or "When will ospreys begin their migration to South America?" Keeping a written phenology helps one analyze cause and effect in nature. By taking notes on temperature, rainfall, and snow accumulation, one can begin to demonstrate the relation of physical factors to changes in populations and growth patterns of plants and animals.

Wood Anemone

Anemone quinquelfolia

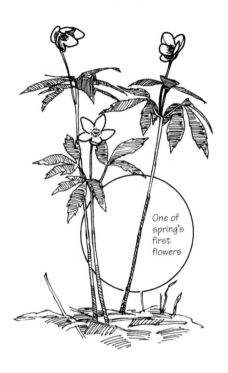

One of
spring's
first
flowers

An early spring flower of open woodlands, the wood anemone has a delicate single blossom that tops a 4 to 8 inch stem and has five (sometimes four) petal-like white sepals that are often tinged with pink-purple. Three long-stemmed leaves branch off in a whorl midway up the main stem into three to five deeply toothed leaflets.

Ancient literature makes many allusions to anemones. Ancient Greeks called the wind *anemos*, and viewed wood anemones as harbingers of spring. In fact, the anemone's common name, windflower, arose from

Pliney's assertion that only the spring wind could open anemones. The Greeks further mythologized the wood anemone through their legend of Venus grieving over the death of her lover Adonis. One poet, in reference to this legend, wrote, "And where a tear has dropped, a windflower blows."

The Persians, on the other hand, adopted the anemone as their symbol of illness because they believed that a wind that passed over a field of anemones was poisoned. The Chinese felt equally ill disposed toward the flower, planting it on graves and calling it the "death flower."

A turn-of-the-century naturalist, said of the solitary bloom that rises before the leaves emerge on the trees, this "delicate blossom . . . holds the very essence of spring and purity in its quivering cup."

A horizontal rootstock anchors the delicate plant in poor soils, and also provides siblings by vegetatively reproducing shoots. The flower opens during the day and closes at night, blooming only a few weeks until it is pollinated by bees and beelike flies. The flower is then shed, and the tiny fruits form rapidly. By early summer, no sign is left that the windflower once braved the unpredictable transformation of a North Woods winter into spring.

Marsh Marigold

Caltha palustris

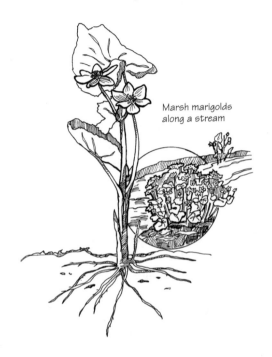

Marsh marigolds along a stream

I n May marsh marigolds carpet wet areas, their brilliant yellow flowers set off by large dark green leaves. The waxy flowers topping long stems have five to nine sepals (no petals) arranged in a shallow cup, creating the overall effect of a large buttercup. The leaves are glossy, roundish to kidney-shaped, and are held up by thick succulent stems.

When blooming the profusion of marsh marigold flowers in wet areas creates a "yellow brick road" effect in the woods. One naturalist wrote that marsh marigolds were so abundant along streams that the ground appeared to be paved with gold. The species name *palustris* is Latin for "of swamps," its primary habitat, while the genus name *Caltha* means "cup."

The budded flowers were collected and sold on the streets of New England every spring in the 1800s along with trailing arbutus. The young leaves were commonly cooked and eaten like spinach in earlier times, but they are poisonous when eaten raw. (It is recommended that the first water be poured off when they're cooked.) As a relative to buttercups, marsh marigolds contain a bitter acid juice that wildlife and livestock avoid.

The Ojibwa used the boiled and mashed roots to heal sores, to prevent childbearing, and to treat coughs because it was said to produce perspiration and loosen phlegm. Frances Densmore, who lived with the Ojibwa in the 1920s and wrote several excellent resource books about their life-style, wrote that the use of the herb was said to be a great secret.

Cowslips, a common name given historically to marsh marigolds, comes from an old word for cow-pies; marsh marigolds were common in wet pastures where cows grazed. Another name, Mary-buds, originated in the Middle Ages, when the blossom was used during church festivals as a devotion to the Virgin Mary.

Blue-bead Lily

Clintonia borealis

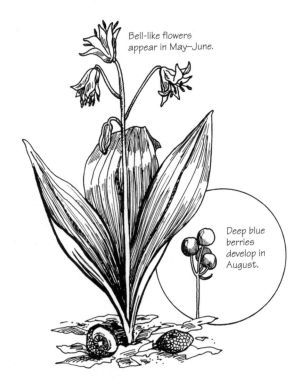

Bell-like flowers appear in May–June.

Deep blue berries develop in August.

The tall (2 feet), leafless, flower stalk carries three to six nodding, bell-like blossoms, rising from three large leaves. The soft yellow-green color of the flowers resembles young corn, giving rise to another common name, corn lily. Each leaf is long (up to 10 inches), glossy spotless green, and narrowly oval with all the veins flowing in an arc to the blunt leaf tip. Deep blue beadlike fruits project from a single upright stalk in August, providing an easy reminder of the common name.

The species name *borealis* trumpets that this is a plant of the North Country. Blue-bead lily blooms from May to June, commonly in untrammeled, shady, moist woods and acid soils. A creeping horizontal rhizome tends to produce dense colonies of blue-bead lilies, much to the delight of any hiker.

DeWitt Clinton, the young governor of New York in the 1830s, and a naturalist and author, was the source of the genus name *Clintonia*. While most politicians have buildings, roads, and ballparks named after them, Clinton received a far kinder accolade when this subtle flower was named for him. Thoreau bemoaned this unholy alliance, writing that politicians were not worthy of such tribute, but the name stood. To make matters worse for the pure of heart, Clintonia is often used as the common name as well.

Chipmunks eat the deep blue berries, and undoubtedly others do also, because the stalks are often stripped by mid-August. However, the bitter berries are considered poisonous to humans. The strikingly beautiful berries should be left to aesthetic, not culinary, tastes.

The Ojibwa used the fresh leaves to treat burns. One botanical source writes that the Ojibwa believed that dogs ate the roots to poison their teeth and thus to kill their prey more quickly.

Hepatica

Hepatica americana

Furry stems

Leathery three-lobed leaves

Hepatica may be the earliest of all upland forest wildflowers. Its flowers of widely varying colors (lavender to purple to pink to white) precede the leaves by many days. The distinctive evergreen leaves are hairy, leathery, and three-lobed, and either rounded or pointed depending on the species. The 3- to 6-inch stem of the flower rises from the ground layer and is quite furry, supporting a cup-shaped flower with five to twelve "petals"—they are really sepals, but only botanists care to note the difference.

Tannin, which can be extracted from the leaves with alcohol, is believed to have some astringent value. Native Americans used hepatica in a host of ways: as a charm to put on traps for fur-bearing animals, and as a cure for vertigo, convulsions, crossed eyes, coughs, and menstrual discomfort.

Hepatica, along with bloodroot, trillium, and violets, has seeds with edible outgrowths called elaiosomes. Ants grab these appendages and drag the seeds back to their hills, where they eat the tasty elaiosomes but not the seeds themselves. They then drag the seeds out of the hill or leave them underground, where some may germinate. Elaiosomes' only known function is as food for ants.

OBSERVATION

Many centuries ago, people searched for medical answers based on what seemed like a reasonable belief—that God would provide physical clues about the use of each plant. Early healers took hepatica's resemblance to a liver as a signal that they should use the plant to treat liver ailments. They were applying the "doctrine of signatures," the belief that the shape of a plant was nature's "signature" regarding its medicinal use. There is no scientific evidence that this system worked.

When first examining a leaf or flower, consider what resemblance, if any, you can find to a part of the human anatomy. Apply the doctrine of signatures, and then be thankful for modern medicine.

Gaywing

Polygala paucifolia

Flowers resemble tiny airplanes.

This unusual orchidlike flower, purple or rose colored, grows low to the ground (2 to 6 inches), arising on a short stem from a whorl of upper leaves. The flowers look like tiny airplanes with two flaring purple-pink wings (sepals), a fuselage made up of petals united into a tube, and a propellerlike bushy fringe at the tip. The leaves are oval, untoothed, and shiny (similar to wintergreen), and grow at the summit of the plant, with tiny scalelike leaves below.

Flowering in May and early June in damp, shady woods, and often in colonies, gaywings look like "a flock of small rose-purple butterflies" that have alighted on the ground, said John Burroughs. This butterfly swarm arises from a creeping horizontal rootstock that sends up the flowering shoots along its length.

The enticing color and structure attracts pollinating bees that land on the flower, forcing the stamens and pistils to be exposed and come into contact with the visitor, thus assuring cross-fertilization. Gaywings produce an underground flower as well, ensuring a seed source in case the aboveground flower fails or bouquet hunters decimate the colonies.

While commonly mistaken for an orchid, gaywings, which are also called fringed polygala, actually belong to the milkwort family. The name polygala is derived from the Greek roots *poly,* meaning "many," and *gala,* meaning "milk." Polygalas were thought to increase the flow of milk in nursing mothers, and so the name. The species name *paucifolia* derives from *pauci,* meaning "few," *folia* meaning "leaves."

Spring Beauty

Claytonia caroliniana

Stems rise from small potatolike tuber

This early spring ephemeral, less than 1 foot tall, usually has two opposite, long oval, deep green leaves halfway up a single stem. Spring beauties are graced with a small cluster of pale pink flowers with deep red veins that give the blossom a peppermint-striped-candy look. The peppermint lines serve as pollen guides that insects follow to the nectaries. Even an apparently all-white spring beauty shows a stripe when seen in ultraviolet light.

The genus name honors John Clayton, an early American botanist. The other *Claytonia* species, *virginica*, seldom reaches into the North Woods, and when it does, is easily distinguished by its longer lance-shaped leaves.

Spring beauty arises from small corms—little round tubers the size of marbles. They can be dug up, boiled for ten minutes, and eaten like a potato, but the amount of food seldom justifies the destruction of the plant unless collected in an emergency.

Gophers and mice dig up and eat the tubers, and elk, moose, and deer graze on the flowers and leaves.

Mrs. William Starr Dana, in her 1893 classic *How to Know the Wild Flowers*, spoke most poetically of the picking of a spring beauty: "What flower . . . is so bashful, so pretty, so flushed with rosy shame, so eager to defend its modesty by closing its blushing petals when carried off by the despoiler."

OBSERVATION

Spring beauty is one of many "spring ephemerals," seeming to come and go in the blink of an eye. The exquisite flowers last only a few days and are soon followed by inconspicuous capsulelike fruits that ripen and fall rapidly. Within a month of flowering, it appears as if the carpet of spring beauties had never been in the woods at all. This early and quick flowering occurs out of necessity: sunlight is a requirement for growth, and however chilly the surrounding spring air may be, the cold temperatures are preferable to the leaf canopy of a mature deciduous forest, which blocks most of the sun by early June.

Bloodroot

Sanguinaria canadensis

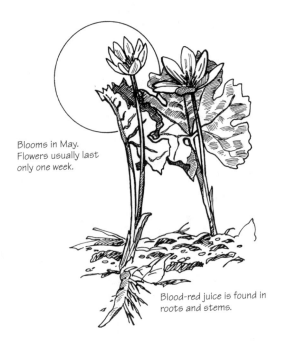

Blooms in May.
Flowers usually last
only one week.

Blood-red juice is found in
roots and stems.

Few flowers can compete with the beauty of the bloodroot's early spring blossom. Usually just one waxy white flower, at first wrapped in a leaf-cloak, rises 6 to 10 inches with eight showy petals (rarely, up to twelve) embellished with numerous golden-orange stamens. The petals expand flat in the morning, become erect by late afternoon, close by evening, and are sensitive to weather changes, closing up when it's cold or cloudy. If picked, the flower almost immediately closes, or the petals drop. The lower part of the stem is sheathed, and the single smooth, large leaf is rounded, with five to nine lobes and a conspicuous cleft at its base. The leaf curves upward and inward like a cupped hand, so the underside with its prominent veins is most easily seen. The 1-inch-long fruit is narrow, single-celled, and elliptical, with many light yellow-brown seeds.

During early May, in forests of maple and basswood trees on floodplains, bloodroot blooms in an ephemeral symphony with trout lilies, hepaticas, and spring beauties. It thrives in the dark-soiled and deeply shaded hardwood understories; it may survive in a clearing for a while, but usually dies when exposed to full sun.

Bloodroot, a perennial that lives in mature forests, requires a stable habitat to reproduce. Stable-habitat plants put little energy into creating fruits attractive to birds and mammals because over time their habitat remains generally uniform. Thus they, unlike annuals and perennials found in pioneer habitats, don't need to produce and distribute masses of seed to help ensure survival of their species. Instead they invest most of their energy in leaf, stem, and root growth to maximize their winter survival into the next spring. Bloodroot has a thick, finger-size rootstock that stores up nutrients in order to rush the flower into bloom in early spring.

Sanguinaria means "bloody," in reference to the blood red juice in the stems and roots. Some Native American tribes rubbed the root juice on their skin as an insect repellent. Others used the red juice to dye their faces or bodies, quills, basketware, and clothing, which gave rise to another common name for bloodroot, Indian paint. The Iroquois used the plant to treat ringworm; a tea from the root was used as a rheumatism remedy by Rappahannocks in Virginia; other tribes used it to induce vomiting. Indians and pioneers cured sore throats by soaking a piece of maple sugar with the root juice and sucking on the lump as a throat lozenge. Bloodroot is considered poisonous—it will cause vomiting, burning in mucous membranes, fainting, and ultimately death, so appreciate the plant's utility only from a historical distance.

ACTIVITY

Carefully pull aside the soil covering the rootstocks of bloodroot, and if many roots branch out, break one of the least substantial roots to see a bloodroot "bleed"—be careful, it stains. Or break a side stem that appears to be of minimal purpose. Cover the roots over again with the set-aside soil and tamp it down gently.

Trout Lily

Erythronium americanum

Trout lily

Trout lily has two smooth, shiny, long, and narrow leaves that are mottled purple, brown, and white—the mottling resembles the marking on a brook trout, hence the name. Usually a solitary light yellow flower with six reflexed petals (actually three sepals and three petals) nods from a single stem. A deeply buried (6 to 15 inches down) smooth white bulb sends out numerous clonal shoots, each shoot soon producing its own new bulb, so trout lily usually grows in extensive colonies, often to the exclusion of other plants.

Trout lily favors hardwood forests or bottomlands, preferring a rich neutral soil. The flowers and fruits are short-lived like most spring ephemerals, needing to get the work of reproduction done before the deciduous canopy unfurls. Each blossom follows the sun, nearly closing at night, and lasts only a few days. By the end of May the flower has usually vanished.

Erythronium is from *erythros*, Greek for "red," apparently in reference to a European relative, or in a color-blind reference to the purplish spots on the leaves.

The very young leaves and the white bulbs can be boiled for ten to twenty minutes and eaten, though the Peterson guide to *Edible Wild Plants* warns that they may be mildly emetic (inducing vomiting). Native Americans used the leaves in tea for stomachaches, hiccoughs, and vomiting, and as a poultice for swelling and ulcers.

OBSERVATION

The seeds may take six or more years to produce flowering plants, so use care not to disturb young plants. Little of the plant shows above ground in the first two years of life, then only one sterile leaf appears for the next four years. Look for large patches of flowerless leaves still in their juvenile years and awaiting their adulthood. Once in adulthood, trout lilies may live three hundred to four hundred years!

Sessile-leaved Bellwort

Uvularia sessilifolia

Bellwort has a single, sheathed stem 4 to 12 inches tall that forks about halfway up, one stem producing only leaves, the other stem producing the solitary, drooping, straw-colored flower. Blooming in May and looking wilted immediately, the upside-down bell-like flower has six long parts that overlap and flare out at the bottom. The narrow leaves, pale underneath, clasp around the stem but aren't pierced by the stem, and are of typical lily appearance, with the veins all in parallel arcs rising to the pointed leaf tip. The flower drops away after one week, and the stem stiffens so much that the plant becomes hard to recognize.

Also called wild oats or merrybells, sessile-leaved bellwort grows commonly in pine woods and thickets, but is absent in most of northern Michigan and eastward. As with so many other northern wildflowers, bellwort reproduces vegetatively by sending up shoots from underground horizontal rhizomes, and may form large colonies. Its three sepals, three petals, and six stamens reveal its woodland lily family heritage. The fruit is a three-winged triangular pod.

The doctrine of signatures (see the glossary) led to bellwort's use as a treatment for sore throats. Someone saw a similarity between the drooping flower and the uvula, the fleshy lobe that hangs at the back of person's throat. The genus name *Uvularia* was derived from this association. Bellwort also was used to lessen skin inflammations and swellings, and Native Americans used a root infusion as an ointment for backache and sore muscles.

The young shoots, sans leaves, may be boiled for ten minutes and eaten as an asparagus substitute, but as with many wild foods, one must seriously consider whether the plant has far better value left alone.

OBSERVATION

Most northern woodland flowers grow less than 18 inches tall. Snow cover in the North Woods winter usually averages about 18 inches, and it's a fair assumption that evolution has favored those plants that hide under a snow blanket rather than stand in the drying winds and extreme cold.

Late Spring
Wildflowers

Barren Strawberry

Wild Strawberry

Oxeye Daisy

Starflower

Large-flowered Trillium

Bunchberry

Columbine

Solomon's Seal

Canada Mayflower

Goldthread

Orchids

Barren Strawberry

Waldsteinia fragarioides

Flowers grow at the top of a leafless stalk.

The bright yellow, five-petaled flowers bear many stamens and appear in May to early June. They often grow profusely in the acid soils of pine forests. Several flowers usually group together at the top of a naked flowering stalk. The leaves look very similar to wild strawberry's, being compound and having three toothed leaflets at the end of a long slender stalk, but are shorter and more wedge-shaped. The big difference between the two plants, evident in the name, is that the barren strawberry lacks runners and strawberries, instead producing a dry, nondescript fruit that no one would bother to pick, much less eat.

A member of the huge rose family, which includes most of the best fruiting species, like blackberry, raspberry, thimbleberry, apple, cherry, and plum, the barren strawberry runs a distant second to the rest of the clan.

Barren strawberry spreads by creeping rootstocks, and holds its flowers often into July. The species name *fragarioides* comes from the Latin root *fraga*, meaning "fragrant."

Wild Strawberry

Fragaria virginiana

This creeping vine has three hairy, dark green leaflets, each with coarse-toothed margins. It produces a white, five-petaled flower. The fruit is much smaller than that of a cultivated strawberry, often barely a bite, but is far superior in flavor. Look for wild strawberries during late June and July in dry open habitats.

Humans, of course, are not the only ones to relish the flavor of strawberries. Thirty-one species of birds and mammals, from cedar waxwings to rabbits to deer, gather strawberries as well.

The leaves make a gentle tea, and the Ojibwa used the boiled roots as a digestive remedy for children. Although wild strawberries make delicious preserves, they may be even better when served simply with a bit of light cream and a sprinkling of sugar. These uses are, of course, dependent on getting the berries home. The very best place to eat strawberries is in the midst of the berry patch where the essence of the warm summer sun seems concentrated in each berry.

ACTIVITY

Pick a wild strawberry and compare its fruit in size and flavor to the domesticated varieties. The small size of the wild strawberry compresses all that wondrous flavor into just one bite. Izaak Walton said all that need ever be said about this fruit, "Doubtless God could have made a better berry, but doubtless God never did."

Oxeye Daisy

Chrysanthemum leucanthemum

Honeybee
seeking pollen.

The leaves of the oxeye daisy are dark green, 2 to 3 inches long, and roughly toothed. The 2-inch-wide daisy flower, a prolific resident of roadsides and open fields, is borne on a two to three foot unbranched stem that waves above its shorter compatriot, hawkweed. Daisies belong to the composite family, named for the fact that each flower is a composite of tiny disk flowers (making up the yellow center) and twenty to thirty ray flowers (the white "petals"). Each "petal" is an individual flower leading to a yellow tubular floret in the center of the eye, containing the pistil and stamens. So, when you pick one daisy, you're actually picking a bouquet.

An escapee from Europe, the oxeye daisy colonizes open fields and roadsides, providing a joyful sight for motorists, but angering farmers who have trouble eradicating it from their pastures. *The Farmer's Encyclopedia* in 1851 referred to daisies as "gawky-looking" and "a blemish." "White weed" and "May weed" were two of its less flattering early nicknames.

Daisies owe their success to a variety of factors. Animals avoid eating the acrid leaves, unwittingly helping to select the daisy as the most successful competitor in a field. Daisies can tolerate poor soil and weather conditions, as well as trampling. Thus they often are a distinct, beautiful sign of a rundown pasture.

Daisies spread so rapidly in part due to their ability to vegetatively reproduce through underground stems that push up to form clones of the parent plant. Mowing a daisy simply encourages it to send more energy into its rhizomes, and so to spread all the more quickly. Seed heads produce hundreds of seeds, which disperse in the wind and are ready to germinate almost immediately.

Daisies are named logically, "the day's eye," from the English daisy, which closes at night and opens at sunrise.

Almost all children have played "he loves me, he loves me not" with the rays of the daisy. If you don't get the desired result on the first try, usually thousands more daisies stand ready to confirm the truth of your love. So, just try again.

Starflower

Trientalis borealis

Starflower

S tarflowers with five to seven snow white petals on thin stems, and seven delicate threadlike stamens rising from the center of each flower, resemble sharp-pointed stars. The long, tapered leaves are also in a star-like whorl below the flower, accenting the blossom. This rather small (3 to 9 inches), deep-woodland flower usually has only one or two fragile-appearing "stars" per plant, but the blossoms often hang on until July.

The name *borealis* refers to the northern boreal forest where starflower is customarily found *(Boreas* was the ancient Greek personification of the north wind). *Trientalis* is Latin for "one-third of a foot," in reference to its height.

After its emergence in early spring, starflower produces no new leaves or flowers. The action for star-flower occurs underground, where by July the plant sends from its root base several rhizomes (underground stems) snaking out just under the forest floor. Nutrients are shunted from the aboveground starflower to the tips of the new rhizomes, each of which begins to swell into a small tuber with an inactive bud. The leaves die in August, and by October there is nothing left to see. The underground connection to the new, genetically identical sister plant decays, and now a few feet apart several starflowers await the spring, when they will emerge through the litter and begin the duplicating process again. Of course, starflowers reproduce through flowers, fruits, and seeds as well.

ACTIVITY

Gently pull back the soil around a starflower to expose the horizontal rhizome, which will produce a clone plant. While humans have ethical concerns about cloning other humans, in the world of plants production of multiple clones ensure reproductive success should seeds fail to be produced or germinate, or, once germinated, die due to extremes of weather, browsing, or disease.

Large-flowered Trillium

Trillium grandiflorum

Single showy flower

With three leaves, three sepals, and three petals, the tall (8 to 14 inches), white flower is so striking, and often so numerous in a rich woods, that at first glance it is hard to see anything else but trilliums. Erect and unbranched, the solitary stem arises from a deeply buried large, coarse bulb that supports a single showy flower. The blossom spreads 2 to 4 inches and is balanced by narrow and shorter dark green sepals that provide a contrasting background. The whorled smooth leaves extend 6 inches or more and are half again as wide, with prominent veins. As the flower ages over several weeks, it may turn pale pink.

Found predominately in mixed hardwood forests with deep, moist, neutral soils, trilliums need a shady undisturbed woodland in which to flourish. Clear-cutting in such a woodland allows too much sun penetration, and the trilliums usually die off.

The name *trillium* derives from the Latin *tri*, meaning "three," since everything, from the leaves to the celled ovary and ribbed berry, comes in threes. Protected by many state statutes against picking, the flower has little fragrance and wilts quickly, making a rather useless bouquet anyway for those who ignore the law.

The fruit, a large (1 inch), angular, green or red or purple, fleshy berry appears in July. Ants distribute the seeds by dragging them back to their nests, where they eat a fleshy attachment to the seed and then discard the seed untouched. Although the seed will germinate the next spring, the trillium won't produce its first blossom for six years or more.

Lake Superior Indian tribes used trilliums as an aid in childbirth, the Potawatomis used the plant for sore breasts, the Ojibwa treated ear complaints and rheumatism with it, and the Menominees found that it reduced eye swellings and cured cramps. Deer eat the leaves, and humans too have in times past eaten the cooked leaves.

OBSERVATION

In June, if you have the patience, wait and watch for ants dragging trillium seeds away to their colonial nest. The seeds of bloodroot, spring beauty, and several violets all bear elaiosomes, which contain certain lipids that the ants can't get from other foods. The ants feed the elaiosome to their larvae. They remove the leftover seed from the nest and dump it in the colony garbage dump, a sort of mini–compost heap that is rich in nutrients and moisture, providing an ideal site for germination. The process of seed dispersal in exchange for an essential food is a form of mutualism, a relationship between two species, in this case, the trillium and the ants, in which both benefit. Without the work of ants, the profusion of trilliums and other wildflowers in the woods would be dramatically reduced.

Bunchberry

Cornus canadensis

June flowering

In late July berries mature.

A common, low (3 to 9 inches) resident of cool, moist open woods and lowlands, the bunchberry is easily recognized by its whorl of four to six leaves on a single stalk, each leaf having parallel veins running from one end of the leaf to the other. Flowering in June, the single white blossom appears to have four petals, but these are really leaflike structures (bracts) surrounding a cluster of tiny greenish yellow flowers in the center. Each of these tiny flowers may become a fruit. The clusters of red berries mature in late July, brightening the ground layer considerably.

The smallest member of the dogwood family, bunchberries often form large, brilliant colonies by cloning themselves through their horizontal, forking rootstocks. Both reproductive strategies, sexual (flowers and fruit) and vegetative (cloning from rootstocks), help to ensure the survival of the species. The flowers and fruits provide new gene combinations, and the rootstocks provide new individuals to produce more fruits and maintain genetic diversity.

Although it doesn't grow in the northernmost regions of the United States, the species name, *canadensis,* describes the bunchberries adaptation to northern habitats.

The raw berries furnished food for the Ojibwa, and are sometimes still eaten today, but they have little taste and large seeds. Songbirds such as the veery and the warbling vireo, along with sharp-tailed grouse, aren't as discerning, and do a Johnny Appleseed job of distributing those large seeds throughout the North Woods.

OBSERVATION

Examine the tiny flowers with a hand lens. Count them to estimate the "bunch" of berries that may form later in summer. Then, go back and check the accuracy of your estimate once the berries have grown.

Columbine

Aquilegia canadensis

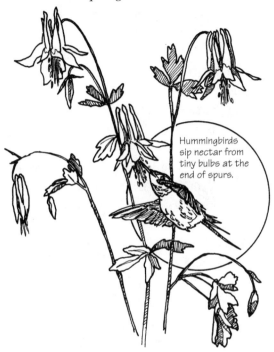

Hummingbirds sip nectar from tiny bulbs at the end of spurs.

The tall-stemmed (2 to 3 feet) flowers have five fused tubular nodding bells that are deep red to pinkish white on the outside and yellow within. Long curved spurs point upward, while the yellow male stamens and female pistils extend and point downwards. The leaves are compound, rounded into lobes, and divided into threes. The June-flowering columbines are slender and graceful, in beautiful contrast to the rocky slopes and poor ground they often occupy.

Some people maintain that the columbine's uniquely shaped tubelike nectaries resemble a circle of doves gathered in a ring around a dish, hence the common name, from *columba*, meaning "dove."

The genus name *Aquilegia* may be from *aquila*, which means "eagle," supposedly referring to the spurs on the top of the nectaries, which are bent like the talons of an eagle.

It is interesting that symbols of peace and ferocity are seen in the same flower. Maybe that reflects the beauty of the flower perched fearlessly in very rugged and harsh habitats.

Hummingbirds and long-tongued hawkmoths sip nectar from the tiny bulbs at the ends of the spurs. Bees that lack the long tongue may nip a hole in the bulb and take nectar the easy way.

The seeds have been crushed and used as a men's cologne, made into an infusion for headaches and fever, and traded whole between tribes as a commodity. In the past, the roots provided a stomach remedy, or when boiled, a food in times of famine. Omaha and Ponca men used columbine as an aphrodisiac; the pulverized seeds were rubbed into the palms, and the suitor would try to contrive a way to shake hands with the woman he desired.

OBSERVATION

Look closely at the unique flower. Do you see a circle of doves, or the curved claws of an eagle, or something very different? Common names often derived from the imaginations of early people. What would you name the flower if it was your choice?

Solomon's Seal

Hairy or True Solomon's Seal *Polygonatum pubescens*
Three-leaved False Solomon's Seal *Smilacina trifolia*
False Solomon's Seal *Smilacina racemosa*

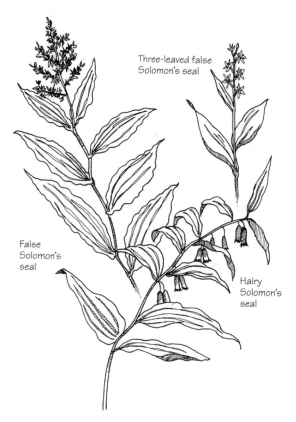

Three-leaved false
Solomon's seal

False
Solomon's
seal

Hairy
Solomon's
seal

These species belong to the lily family and have the long, parallel-veined leaf typical of lilies. All their flower parts grow in threes or multiples of three. Hairy and false Solomon's seal look somewhat alike in over-all structure but the bell-like flowers in the hairy Solomon's seal hang like Christmas tree ornaments, usually singly or in pairs from the axils of the leaves. The false Solomon's seal's tiny star-shaped flowers bloom in a large, branched, pyramidal cluster on a stem rising above the leaves. The three-leaved false Solomon's seal bears little resemblance to the previous two species—it's much shorter (4 to 10 inches) and weaker. It grows almost exclusively in wetlands or bogs; the leaves are all "sheathed" to the main stem; and the white star-shaped flowers number only a few around a stem reaching above the plant.

The origin of the name Solomon's seal remains in some dispute. According to one theory, it refers to the large scars on the rootstock, which were thought to resemble King Solomon's official seal for legal documents. According to the second theory, the name comes from the plant's use as a balm, or seal, to close fresh wounds. The third holds that the name originated due to the resemblance of the six-petaled flowers, and the six points on each blossom, to the six points of the Star of David.

The hairy Solomon's seal grows in rich forest loam, usually disdaining the sandy soils that characterize much of the North Woods. The single, arching 3-foot-tall stem rises from a jointed and scarred rhizome—the genus name *Polygonatum* refers to the rhizome's resemblance to a series of knee joints, *poly* meaning "many" and *gonu* meaning "knee or joint."

The yellow-green or greenish white flowers are inconspicuous, the sepals and petals united into a narrow tube and hanging on slender stems under the leaves, with the blue-black fruit following in late summer.

The juice from the crushed rhizome has been used for earaches and sunburn, and a tea from the leaves was brewed for a contraceptive. Native Americans and settlers ate the starchy rhizomes, though one source says the colonists ate it only to avoid starvation (not much of a testimonial). Current literature recommends that it be boiled for twenty minutes and served like potatoes. The new shoots can be boiled for ten minutes and supposedly taste like asparagus. As with all wild foods, one must balance culinary zeal with appropriate ecological restraint.

False Solomon's seal grows commonly throughout the North Woods, content in dry or moist woods, and often next to true Solomon's seal. Also known as Solomon's zigzag, it has a single stem, arched, coarse, unbranched, and meandering in its growth. The rhizome lacks the leaf scars of true Solomon's seal.

In late summer, at the tip of the stem a cluster of aromatic pale red berries, often speckled with purple, ripens. Birds such as those in the thrush family relish the berries. People sometimes eat them, but they are reported to be mildly cathartic. Another source, though, writes that Native Americans ate the "delicious" berries in large quantities and that settlers ate them for their Vitamin C and called them scurvy-berries. The young shoots can be chopped and added to a salad. The Ojibwa boiled the roots after first soaking them in lye to eliminate the bitter taste.

This species has been saddled with an unfortunate name, "False" Solomon's seal, which suggests that it is lacking in something. In 1895, Schuyler Matthews wrote in defense of its beauty that we "may as well call a Frenchman a false Englishman."

Three-leaved false Solomon's seal, a name much larger than the plant, requires cool, acid soils, and thus is usually seen only by those who enjoy bog trotting. Its extensive horizontal creeping rootstock produces new clonal shoots, a rather typical adaptation of bog plants for reproducing in such a difficult habitat. The white star-shaped flowers in slender clusters blossom on a terminal stem in June, and dark red berries appear from late July to September. *Trifolia* means "three leaves" and refers to the three long and narrow leaves that are sheathed to the stem.

Canada Mayflower

Maianthemum canadense

Fruit appears in July.

Canada mayflower grows low to the ground (2 to 6 inches), blooming from early May into June in a cluster of small white flowers on a common stalk. They are usually too small and insignificant to be picked for bouquets even though the blossoms are wonderfully fragrant. Two or three shiny leaves unfurl in early May, each heart-shaped at the base, pointed at the tip, and ruffled in between, with veins all emanating from the base and arching to the tip in typical lily fashion. Sometimes only one long-stemmed leaf will appear, and no flower. The flowering stem often zigzags a bit on its way up. The fruits initially appear as clusters of speckled berries in July and gradually turn pale red by September, often lasting into the fall and turning wine red, in lovely contrast to the yellow-brown leaves.

The Canada mayflower qualifies as a ubiquitous North Woods wildflower. It grows in sandy woodlands, but also in a variety of habitats from sands to cedar swamps. The threadlike horizontal rhizomes create clonal colonies of mayflowers that profusely dot the ground, making it sometimes difficult to walk without crushing a few.

Don't be fooled by the name, for this wildflower is found throughout the North Woods and south and does not limit its range to Canada. Canada mayflower is also called "wild lily-of-the-valley" because of the resemblance of its leaves to the domestic variety and the similar scent of both flowers. Beadruby is yet another common name, in obvious reference to the red fruits. *Maianthemum* is Greek for "Mayflower" from *maios*, "May" and *anthemon,* "blossom."

Native Americans reportedly used the rootstock for medicinal purposes and the berries for food, though no current references indicate that the fruits are edible by humans. Canada mayflower offers rather limited wildlife value—grouse, chipmunks, mice, and some ground-feeding birds eat the fruits, and snowshoe hares consume the whole plant.

ACTIVITY

To see just how densely Canada mayflower may grow, count how many you find in a 2-by-2-foot square. Try different habitats and compare densities.

Goldthread

Coptis trifolia or *groenlandica*

Look for yellow underground rhizomes.

The delicate flowers that appear in May, grow singly on long stems with five to seven, white petal-like sepals arranged in a star. Fifteen to twenty-five white stamens with gold anthers, and numerous pistils and filamentlike petals (that are really nectaries) give the center of the flower a furry appearance. Emerging from the base of the plant, the upright lustrous leaves rise on 3- to 6-inch-long stems, each evergreen leaf divided into three fan-shaped, scalloped leaflets. The leaves look a bit like those of the barren strawberry. If you are in doubt about its identification, carefully expose the slender, brilliant yellow horizontal roots from which the name was derived.

Goldthread thrives in cool, moist woods, particularly in cedar swamps, and in bogs. Well adapted to acid soil, it usually grows under conifers, and often in moss. The rhizomes vegetatively reproduce by sending up shoots, often creating carpets of goldthread in deep woods.

Goldthread was once called "canker-root"; the brilliant yellow underground rhizomes have astringent properties and were used as a remedy for sore and ulcerated mouths. Mohegans and Montagnais boiled the root and used the solution for a gargle; Penobscots and others chewed the stems to prevent sores in the mouth. The root was also used for lessening the pain of teething. The root contains the alkaloid berberine, which produces a mild sedative. Widely used as a folk remedy, goldthread roots dried for market in 1908 fetched sixty to seventy cents a pound.

Tea brewed from the golden root was also used as a bitter tonic in the spring, undoubtedly because of the belief that anything that tasted bad must be good for you.

The genus name *Coptis* is from the Greek, "to cut," in reference to the cut-leaf; the species name *trifolia* means "three-leaved." The other accepted species name, *groenlandica*, refers to the northern habitats in which goldthread resides.

OBSERVATION
Scrape soil away from the roots to expose the wiry rhizomes "made of gold." Carefully replace the soil and pat it down.

Orchids

Sweet fragrance lures insects inside pouch.

Yellow lady's slipper

No other plant family combines such remarkable fragrances, tropical colors, and unique configurations as orchids, and many of the common names reflect their exotic status: dragon's-mouth, ram's-head lady slipper, calypso, and rattlesnake-plantain. Orchids arrange their flower parts in threes (three sepals, three petals), but the outstanding feature of all orchid flowers is the one petal that has been modified into a "lip." This lip may look like a fringed tongue (rose pogonia) or like a tiny moccasin for elves (pink lady's slipper). But most importantly it serves the function of a "landing pad," luring insects to perform the vital function of cross-fertilization.

Not all orchids dazzle the senses. Many are small and inconspicuous (like Loesel's twayblade or checkered rattlesnake-plantain), identifiable as orchids only by the use of a hand lens. Others defy initial attempts to place them with the orchids, like spotted coralroot, which has no green leaves and is saprophytic, looking more like some odd tall fungus than an orchid.

The dustlike seeds from orchids can ride on the wind for many miles, but since they contain very little or no stored food, they must germinate rapidly in just the right conditions and most often also require that a specific fungi be present in the soil. The symbiotic marriage of the two, fungus and orchid, appears to help the orchid absorb nutrients from the soil throughout its life. And its life may be long indeed, some species taking more than a dozen years to flower and produce seed.

The combination of exacting habitat conditions and slow flowering makes orchids very difficult to transplant, and conservation laws in most states protect all orchids from removal. The most serious threat, however, to our orchids is not bouquet pickers, but habitat destruction. Drain a white cedar swamp or bog, and as a result, a host of northern orchids are likely to disappear.

Yellow lady's slipper (*Cypripedium calceolus*), is one of the most common northern orchids because it grows in a variety of undisturbed habitats. Its lustrous coloration and slight but sweet fragrance attract insects. The dark interior pouch is made less forbidding by "windows" on the walls of the slipper that let light in and guide the insect to the right spot for pollination. Insects fly from one lady's slipper to the next pollinating each as they gather sweet nectar. Studies show that it may take sixteen years after the seeds germinate for the yellow lady's slipper to reach maturity and produce flowers.

The sandy pine woods throughout much of the North Woods provide the best habitat for pink lady's slipper (*Cypripedium acaule*), though they are common in bogs and mossy hummocks as well. The seemingly dissimilar habitats have two things in common—both are acidic and both lack nitrogen. In fact, nitrogen appears to be toxic to some orchids.

The lady's slipper had many historical medicinal uses. The Ojibwa and Menominees used a root decoction for "female problems," while the Penobscots made a tea to calm the nerves. American medicine in the nineteenth century employed the lady's slipper as a drug for insomnia, as a sedative for the nerves in general, and as an analgesic (pain reliever).

The quaking bog mats of the North Woods offer the adventurous hiker a June reward of grass pink (*Calopogon pulchellus*) and rose pogonia (*Pogonia ophioglossoides*). Grass pink has a yellow-crested lip held above the rest of the petals, unlike other orchids, whose "lip" hangs below the petals. The common name derived from the color and the single, slender, grasslike leaf.

Rose pogonia grows alone on a single stalk, with pink petals and sepals providing the background for the beautiful yellow bearded lip. One lance-shaped leaf clasps the stem halfway up, and another leaflike bract clasps the stem just below the flower. The violetlike fragrance helps to attract insects.

Coralroot orchids are found in both deciduous and evergreen woods. Four species grow in Wisconsin; each has underground roots that are branched like ocean corals and lack chlorophyll. Coralroots, the only North Woods orchids that do not have green leaves, must obtain nutrients from dead organic matter in the soil, much as mushrooms do. The erect stem of the spotted coralroot (*Corallorbiza maculata*), the most common species in the sandy pine habitat near my home, bears many dull purple flowers, each with a white lip spotted with red.

OBSERVATION

To see dozens of pink lady's slippers in both a bog habitat and upland sandy habitat, try the Trout Lake nature trail in northern Wisconsin, within the Northern Highlands State Forest, from early to mid-June. A lengthy boardwalk allows you access to a bog rich in lady's slippers, and the upland trail leading there has many as well. Or get out and discover a bog walk where ever you live.

Early Summer Wildflowers

Orange Hawkweed

Common Milkweed

Wintergreen

Sarsaparilla

Black-eyed Susan

Indian Pipe

Pipsissewa

Twinflower

Orange Hawkweed

Hieracium aurantiacum

Livestock don't eat orange hawkweed.

A s a member of the composite, or daisy, family, each orange hawkweed flower is really a bouquet of flowers, composed of a yellow disk flower in the center and many orange to red ray flowers radiating out from the disk. The leaves rise in a distinctively hairy rosette at the base of the plant, and the leafless stem is hairy as well. The black hairs led gardeners in the 1500s to name it "Grim the Collier" because it looked as if coal dust had been shaken onto the plant. Orange hawkweed itself is not a native North Woods species, but two other less prevalent hawkweeds are native to the North Woods.

The name hawkweed is apparently derived from a folktale that hawks ate this flower to aid their vision. Even the genus name *Hieracium* derives from the Greek *hieros,* for "hawk." Naturally the doctrine of signatures (see the glossary) led people to use hawkweed as an herbal remedy for poor eyesight, but with little apparent success.

Hawkweed is often referred to as the "devil's paintbrush" because it invades farmer's fields and spreads quickly. Livestock don't eat it, so it's indirectly selected for survival in farm fields since nearly all the other plants are grazed down. Its presence indicates poor farm soils, and one means of controlling it is by fertilizing the soil to produce rival plants that will crowd it out.

OBSERVATION

Note the colorful profusion of yellow and orange hawkweed, interspersed with oxeye daisy, along our roads in midsummer. Orange hawkweed exudes a toxin to inhibit other plants around it, giving itself an exceptional competitive edge in the already hostile, sandy, hot environment in which it thrives. Add to that advantage its ability to shade out other ground plants and it's no wonder that roadsides are often clothed in hawkweeds.

Common Milkweed

Asclepias syriaca

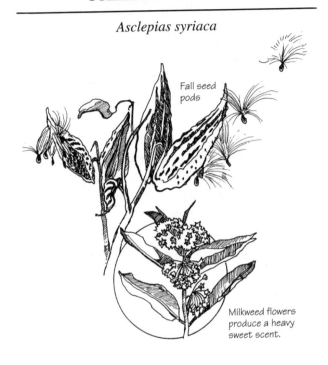

Fall seed pods

Milkweed flowers produce a heavy sweet scent.

Tall plants (3 to 5 feet), with opposite, oval, large (up to 9 inches long and 4 inches wide) leaves, milkweed produces spherical clusters of 1/2-inch-diameter pink to purple flowers that often droop with the weight of the blossoms. An easy aid in identification is the milky and very sticky sap that oozes from the stem and leaves when they are broken. The seed pod is long, pointed, and warty, and holds thousands of silky white hairs, with a single seed attached to one end, that sail effortlessly in any breeze. Many species of milkweed live in the North Woods, but the common milkweed is, in keeping with its name, the most prevalent, and is found along roadsides, in wastelands, and in dry fields.

If the seemingly lighter-than-air seeds aren't enough to distribute this plant, milkweed also spreads by horizontal root suckers up to 15 feet long that anchor the soil, ensuring its roadside presence for years to come.

In July, a patch of flowering milkweed offers a nearly overwhelming fragrance. The heavy sweet smell, comparable to the power of a lilac hedge, is very attractive to a host of butterflies, bees, moths, flies, and other insects. Milkweed is well known as the plant on which monarch butterflies deposit their eggs. When an egg hatches, the caterpillar eats the milkweed leaves for about two weeks and then attaches itself to the underside of a leaf, forming a chrysalis. The adult monarch butterfly emerges in late summer and then leaves on its annual migration to central Mexico.

The milkweed leaves contain cardiac glycosides, which when eaten by the caterpillars and adult monarchs makes them toxic to birds and other predators. Most predators have therefore learned over time to leave the monarchs alone.

Humans have found many uses for milkweed. The genus name, derived from *Aesculapius*, the Greek god of medicine, indicates milkweed's extensive historical medicinal value. But nonmedicinal uses were common too. During World War I, children who collected the silky seeds for stuffing life preservers, were paid a penny a pound for their efforts. Early settlers used the silk for pillows and mattresses. The Dakota used them to staunch wounds. Weavers have historically mixed the silk with their wool or flax to lighten the fabric.

ACTIVITY

1. Break a lateral stem from a milkweed and touch the "milk." The sticky white sap served as a glue for settlers and contains latex, a rubberlike material.

2. In August, watch for the formation of a monarch chrysalis on the underside of the leaves. The chrysalis may be removed carefully and brought into a school classroom allowing children to watch the eventual emergence of the adult monarch, an event that usually occurs within a few weeks of the chrysalis formation.

Wintergreen

Gaulthers procumbens

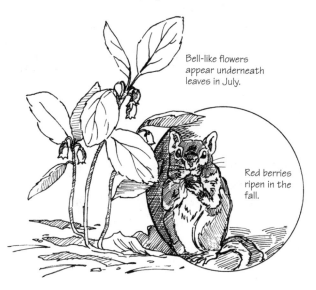

Bell-like flowers appear underneath leaves in July.

Red berries ripen in the fall.

Wintergreen has a low to the ground (2 to 6 inches), creeping woody stem bearing three green and glossy oval leaves that exude the wintergreen fragrance when crushed. In July, waxy, bell-like white flowers with five teeth hang hidden beneath the leaves. The leaves are evergreen, and appear light green when young, but turn dark green and leathery as they grow older. The round, red five-celled fruits ripen in the fall and remain attached to the plant all winter. Their taste is the distinctive delightful wintergreen flavor.

Wintergreen vegetatively reproduces by an underground stem that sends up shoots. The evergreen leaves, waxy and waterproof, retain water and so are well adapted to the desertlike environment of winter.

Deer, bear, grouse, chipmunks, and mice eat the nutritious berries and often the leaves, which are likely a very welcome sight to wildlife after the snow melts in April. Hikers also partake of either berries or leaves; and a particularly good combination is blueberries and wintergreen berries. The flowers also taste of wintergreen. Wintergreen may be found in all forests, but is most common in conifer stands.

The Ojibwa made the leaves into a beverage, commonly boiling water whose purity they were unsure of and flavoring it with plants like wintergreen, black cherry, and Labrador tea.

Wintergreen oil can be distilled from the leaves, but it takes one ton of leaves to produce one pound of oil, an effort hardly worth the trouble and the harm to the woods. Wintergreen oil was also historically distilled from the leaves, bark, twigs, and roots of yellow birch, requiring one hundred saplings per quart of oil. These days commercial wintergreen oil is synthetic.

ACTIVITY

An excellent tea can be made from the dried leaves. Steeping with fresh leaves yields only a weak tea unless the mixture is allowed to sit for a day.

Sarsaparilla

Aralia nudicaulis

A single, tall (1 to 2 feet) leafstalk bears a single compound leaf, which branches off into three large leaflets, each with three to five leaves. The branching pattern is akin to bracken fern and provides a key clue in identifying sarsaparilla. The very small white-green flowers usually bloom in three round clusters in late June, rising inconspicuously on a single short leafless stalk that is well hidden under the large leaflets. Purple-black berries form in late July in similar round clusters beneath the large leaves. The leaves turn bronze in the fall.

Wild sarsaparilla grows commonly throughout open moist woodlands, spreading through thick underground roots running three feet or more to form colonies of clones. I have yet to find any flavor in the roots even after long boilings, though most books will tell you that the wild plant can be used as a substitute for the soft drink flavoring: the true flavorful sarsaparilla, not our native one, comes from a small vining bush found in Mexico and the Caribbean.

Found in the ginseng family, sarsaparilla has a wealth of historical medicinal uses. The Ojibwa used the root as a remedy for "humor in the blood," as a treatment for sores like boils and carbuncles, to stop women's periods, and, combined with the root of calamus in a decoction, as a charm to be put on fish nets to attract fish and to "rattle snakes away." The Penobscots combined the roots with sweet flag for a cough remedy, the Potawatomis made a poultice from the roots for swellings and infections, and the Catawbas made a tea from the roots to be used as a general healthful drink. One New England botanist claimed that the Indians consumed the roots during wars and hunts because of their sustaining value, while yet another source says the root was used to relieve insect bites.

A host of birds such as white-throated sparrow and wood thrush, consume the berries, spreading the seeds to locations throughout the eastern United States. Mammals like black bears, foxes, skunks, and chipmunks eat the fruits as well. Some sources say burrowing rabbits eat the fragrant roots.

OBSERVATION

Uncover the roots of a sarsaparilla in a patch and follow the horizontal roots that lie just under the soil. Just about every sarsaparilla is "holding hands" underground, linked in a genetic chain to an original plant. Replace the soil carefully over the roots and tamp down.

Black-eyed Susan

Rudbeckia hirta

A host of insects pollinate black-eyed Susan.

Tall (1 to 3 feet) with rough, hairy stems and leaves, the familiar blossoms have eight to twenty yellow to orange "petals" (ray flowers) surrounding a rounded purple-brown "eye" (disk flowers). The center eye is rarely black, so the common name ought to be brown-eyed Susan. The rays curl backward over time, forming a cone, hence the other common name, coneflower.

The black-eyed Susan has sterile rays. The "eye" holds hundreds of tiny flowers packed together to create a "disk," each little tube-shaped floret having male and female parts. Usually the disk opens in stages: the center has unopened disk flowers; further out a circle of male flowers open, producing pollen; and the outermost edges appear fuzzy, with open female parts ready to receive the pollen. When a bee lands on the edge of a disk flower, it pollinates the outer circle of female flowers and, as it works its way toward the center, picks up the male pollen, eventually reaching the unopened center, signaling the bee to move to the next flower, where it begins the cross-fertilization process again.

A member of the dry roadside and open field community of flowers, black-eyed Susans light up many a traveler's journey in mid to late summer. Unlike its common companion, the alien oxeye daisy, black-eyed Susans are native to prairies in the United States, though not to the North Woods. They arrived from the western clover fields in hay bales (it's remarkable how much of our present flora is cow-inspired), and spread rapidly throughout the North and East. Some black-eyed Susans are biennials—a rosette of leaves forms in the first year only, and the flower forms in the second year— while others are perennials.

The rough, hairy stems are very likely a water-conserving adaptation for the dry habitats black-eyed Susans frequent. A tough, widespreading root system helps the plant grow in large colonies that may encroach into agricultural fields where it is heartily disliked by farmers.

The sterile yellow ray flowers and dark eye draw in a host of insects (bees, wasps, butterflies, and beetles), ensuring cross-fertilization and genetic vigor. The dark "eye" may be yellow with pollen during fertilization.

Indian Pipe

Monotropa uniflora

In late summer, the pipe rises.

The white, waxy-looking 3- to 8-inch stem, scalelike leaves, and nodding flower of Indian pipe are just about unmistakable. With no greenery whatsoever, it looks more like a fungus than a woodland flower. Ghost plant is another common name, derived from the translucent white color of its stem and flower. In late summer the nodding pipe rises upwards, and the plant turns black and begins to decompose (when picked, it also quickly blackens), showing the origin of yet another common name, corpse plant. The fruit forms into a capsule that splits down the sides, releasing fine brown seeds.

Indian pipe is found in virtually all coniferous habitats, from rich shady woodlands to bogs to dry sandy pine forests.

Without chlorophyll and unable to photosynthesize, Indian pipe has to obtain nutrients another way. It does so by sharing the mycorrhizal fungi that are attached to the roots of conifers. These fungi get their nutrients from the tree's roots, and in exchange help extend the tree's root system out further into the soil. The mycorrhizal fungi are also attached to the roots of Indian pipe, which, acting almost as a parasite on the fungi, receives its nutrients from the fungi. Researchers discovered this fact by injecting radioactive carbon into the bark of a spruce; five days later nearby Indian pipes had already become radioactive. So while Indian pipe is not a direct parasite of conifer roots, it does obtain its nutrients indirectly through them. Indian pipe may also receive some of its nutrients from the decay of dead organisms in the soil.

Native Americans and settlers bruised the plant and applied the clear fluid of the stem as an eye lotion.

Montropa, the genus name, means "one turn," in reference to the rotation of the flower from nodding downward to pointing skyward. *Uniflora* has an obvious definition: *uni* means "one," *flora* means "flower."

OBSERVATION
Watch the flower change color over time as it fruits and dies, turning from translucent white to deep black.

Pipsissewa

Chimaphila umbellata

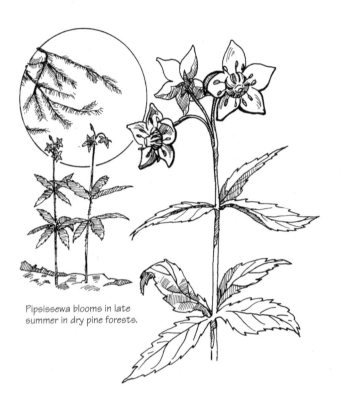

Pipsissewa blooms in late
summer in dry pine forests.

Pipsissewa is most easily identified by when
and where it blooms—mid-July in dry pine forests.
In the deep woods very few flowers just begin to blossom so late in summer. To be sure of your identification, look for the flowers, colored a soft cream with a
pink center ring, nodding toward the earth from a long,
slender, reddish brown stalk. Five petals with violet anthers encircle a large green ovary and broad style that
is sticky on its end. In its center, the flower is waxy and

dainty, with a gentle scent. The waxy evergreen leaves with sharp sawtooth edges form a whorl around the stem.

Pipsissewa vegetatively reproduces through horizontal rootstocks just under the surface.

The name pipsissewa originated from the Cree Indian word *pipisisikweu,* meaning "it breaks it into small pieces." The Crees apparently used a decoction from the leaves to break down gallstones and kidney stones, thus the name.

Michael Weiner, in his book *Earth Medicine, Earth Food,* wrote that the Mohegans and Penobscots, Indians of the Northeast, used pipsissewa "to draw out blisters . . . (they) steeped the plant in warm water and applied the liquid externally. . . The Thompson Indians of British Columbia pulverized the entire plant when fresh and applied the mass in a wet dressing to swelling of the lower legs and feet." Other tribes used the leaves to induce sweating and to treat rheumatism and backache, and pipsissewa was officially listed in the *United States Pharmacopoeia* from 1820 to 1916 for its astringent, or tissue-drying, properties.

The Fieldbook of Natural History by E. Laurence Palmer, the best source I know for thumbnail sketches of all North American plants and animals, indicates that the leaves have a delicate flavor and are often nibbled by woodsmen, while other sources claim the leaves have a wintergreen flavor and a tea can be made by steeping them in hot water. But I must respectfully disagree—I skeptically tried a few leaves from pipsissewa and they were too bitter to eat. The bitter taste may keep this plant, relatively uncommon in sandy northern forests, from being depleted.

The genus name *Chimaphilla* means "a lover of winter," probably in reference to its evergreen adaptation.

OBSERVATION

Use a hand lens to really see the pink center ring and the violet anthers—this is a very beautiful flower. Also dig gently around the plant until you find the horizontal rhizome running over to the next pipsissewa, though which is the parent is anyone's guess, since size offers no clue.

Twinflower

Linnaea borealis

Twinflower is a rather inconspicuous low-trailing plant with a pair ("twin sisters") of nodding, pink, funnel-shaped flowers that give off a wonderful fragrance. An evergreen, twinflower has nearly round, opposite leaves with sparse teeth and short petioles (stalks). The stem usually creeps along the shady northern forest floor.

The species name, *borealis*, describes the twinflower's range: northern boreal forests (rarely south of the tension zone). The flowers appear in late June, often forming carpets of delicate hanging bells in a surprising variety of habitats from dry woods to bog edges.

The genus name *Linnaea* refers to Carl Linnaeus, who was portrayed in paintings with a sprig of twinflower in his buttonhole. Linnaeus is the eighteenth century Swedish naturalist who established the binomial system of naming plants—one name for the genus and one name for the species. Prior to Linneaus, plants were given a single name, generously supplemented with descriptive nouns and adjectives. Linneaus' *Species Plantarum*, published in 1753, established rules for nomenclature that clarified the plant classification system. To him we owe many of the tongue-twisting Latin names that, while often unpronounceable to non-botanists, allow us to identify specific plants amidst a now even more bewildering array of common names.

Grouse and deer eat twinflowers in the North Woods, as do mule deer and bighorn sheep out West.

ACTIVITY

Linnaeus consented to have twinflower named after him. If you could pick one flower to name after you, which would you pick? What characteristics would that flower exhibit that speak to your personality? Would you choose personal values such as beauty, fragrance, and rarity, or ecological values such as enriching the soil or providing food, cover, and nesting and denning sites for wildlife?

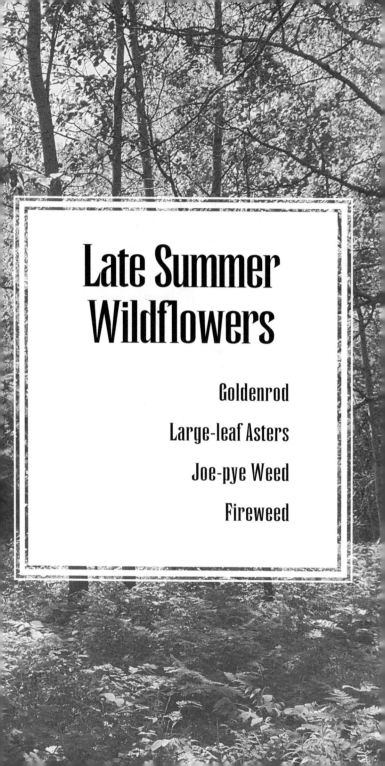

Late Summer Wildflowers

Goldenrod

Large-leaf Asters

Joe-pye Weed

Fireweed

Goldenrod

Solidago spp.

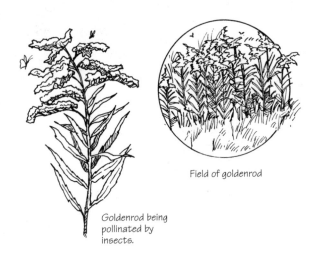

Field of goldenrod

Goldenrod being
pollinated by
insects.

Most everyone recognizes goldenrods, but exact
species identification is difficult, with dozens of
species found in the North Woods. The radiant yellow
of the tiny ray flowers is concentrated in showy clusters atop slender stems 1 to 6 feet tall. Leaves are most
often long, narrow, and toothed, but variations among
species make generalizations unreliable.

While many of our roadside flowers are foreign
imports, goldenrod is a native. Goldenrods spread
through underground rhizomes, which send up new
shoots every year. Colonies of clonal goldenrod can
become very dense and large—some are estimated to
be one hundred years old. In the winter, goldenrods
commonly have wasp galls on their stems. Insects
cause these swellings by laying their eggs inside the
stems. In an attempt to heal itself, the stem responds

to the egg mass by forming a growth around the eggs that protects them through the long winter.

The reputed healing powers of goldenrods inspired the genus name *Solidago*, meaning in Latin "to make whole." Ojibwa called it the "sun medicine," and used it for fevers, sore throats, chest pains, and other ailments. Native Americans and settlers used the flowers for a natural dye. The leaves of several species contain latex which could be used as a rubber substitute if needed.

Hay fever sufferers often blame the flower plumes of goldenrod for their late summer ills, but the blame belongs mostly with ragweed. Goldenrod is insect pollinated, not wind pollinated like ragweed. Insect pollinated plants send very little pollen into the air because their pollen is too heavy. Their brightly colored flowers attract insects, and the nectar-seeking insects inadvertently carry the pollen to other plants. Various bees, flies, beetles, spiders, and butterflies spend a lot of time in goldenrods accomplishing this task. Late-summer sneezing is not the fault of goldenrod, which has become the scapegoat of the plant world, but which is really just the victim of unfortunate timing.

Large-leaf Aster

Aster macrophyllus

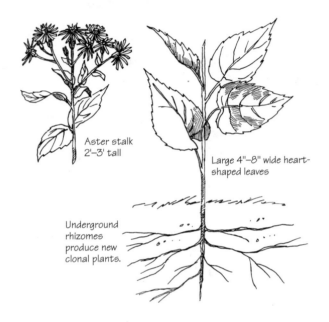

Aster stalk 2'–3' tall

Large 4"–8" wide heart-shaped leaves

Underground rhizomes produce new clonal plants.

The North Woods contain dozens of aster species, varying greatly in color, size, and habitat. But the most common aster in relatively open woods or shady dry areas is the large-leaf aster, which often forms dense colonies that exclude virtually every other type of plant. The rough leaves are quite large (4 to 8 inches wide), rounded, heart-shaped with a notch in the base, and long stemmed. Often sparse, the flowers appear very late in the summer.

All asters belong to the composite, or daisy, family, and exhibit the characteristic cluster of yellow disk flowers in the center and a spray of white to purple ray flowers forming the "petals." The large-leaf aster sends its purplish flowers up on a stout 2- to 3-foot stalk bearing small oblong leaves.

Large-leaf aster assures its survival in a variety of very effective ways: by sending out long underground stems that produce new clonal plants away from the parent plant; by producing a toxin from its roots that weakens competitors; and by creating a subcanopy of large leaves above the forest floor that shades out most competitors—much like the bracken fern.

Why does aster bloom so late in the year? One possibility is that during late summer there are few competitors fighting for the available sunlight and soil nutrients. Over 70 percent of the North Woods wildflowers bloom by June 15. On the other hand, autumn days are shorter and cooler which would seem to offset the competitive advantages.

Aster, Greek for "star," is the source of many celestial words such as astronaut, astrology, astronomy, asterisk (the "little star"), and even disaster ("ill-starred").

The Ojibwa boiled the young, tender leaves of the large-leaf aster along with fish. Early settlers supposedly used the large coarse leaves as toilet paper, thus bringing to life the off-color (but easily remembered) nickname of "large-ass leafster."

Joe-pye Weed

Equpaorium maculatum

Red admiral butterfly
pollinating joe-pye
weed

This very tall (up to 10 feet, but usually 3 to 5 feet),
late-summer flower, most commonly grows leaves
in a whorl of three to six. The leaves are saw-toothed
and large, up to 12 inches long. The delicately fuzzy,
trumpet-shaped flowers form a dome-shaped mass,
and exhibit pink-purple colors (some refer to them

as wine-stained) that are attractive to humans and insects alike. The purple stalk remains through the winter, but the small dry fruits and tiny parachuted seeds disperse in the autumn winds. The crushed leaves have a sweet vanilla odor.

Look for the magenta flowers in August along roadsides with poor drainage and along the transitional edges between wetlands and uplands. As with so many North Woods plants, joe-pye weed also vegetatively reproduces itself through horizontal rhizomes spreading just underground.

Joe-pye weed is the only plant I know whose common name relates to a real person (though many Latin scientific names contain people's names). Joe Pye was an Indian skilled in healing with herbs who lived in rural New England during the late 1700s. His particular talent was reducing fevers, so Native Americans and settlers called their medicinal plant for fevers and kidney complaints the joe-pye weed.

Various tribes believed joe-pye weed held different medicinal properties. The Flambeau Ojibwa used a lukewarm decoction as a wash for sore joints. They bathed children with it to strengthen them, and to help them sleep when fretful; the Potawatomis used it for burns and to clean a woman's reproductive tract after childbirth; the Menominees used it for urinary tract ailments; and the Meskwakis found it to be a love medicine.

The genus name *Eupatorium* refers to the herb doctor Eupator, who in biblical times used this plant. Today, though, joe-pye weed has no recognized medicinal value.

Joe-pye Weed

OBSERVATION
Note that the whorled leaves don't line up with the whorls of leaves above or below, an adaptation that allows sunlight to reach each layer of leaves.

Fireweed

Epilobium angustifolium

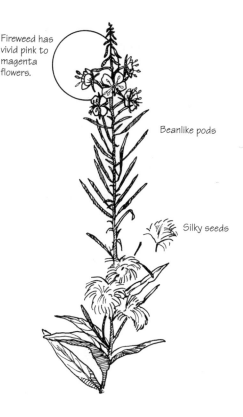

Fireweed has vivid pink to magenta flowers.

Beanlike pods

Silky seeds

The slender spikes of vivid pink to magenta flowers have four rounded petals with four darker, slender sepals underneath which support the petals. The reddish stems are 3 to 6 feet tall with alternate, toothless, narrow, pointed leaves. Below the open flowers, the long beanlike seedpods form in the late summer. New flower buds continue to form above the blooming flowers and droop down as they await their turn to open.

So on the fireweed several stages of flower and fruit development can be seen at any one moment.

You may have noticed light silky seeds being carried on the wind in late summer. These are the seeds of fireweed, and their pervasiveness reminds me of the spring "seedstorms" of aspen, cottonwood, and dandelion. The seedpods, which angle upward on the stem, split down the seam, allowing the seeds to join the wind in search of an open area.

Fireweed is a pioneer species springing up on disturbed sites, particularly on burned-over areas (which may be the origin of its name) or along logging roads. It was the first plant to colonize the rubble of London after the bombing of World War II. Fireweed may also earn its name from the profusion of magenta blossoms that often light up a roadside or field, looking a bit like a fire of flowers.

The Ojibwa used the leaves as a poultice for bruises. Today, beekeepers appreciate the flowers as a consistent source of nectar. The young shoots and leaves are said to be a substitute for asparagus, the leaves may be used as a tea, and the pith of the stems may be placed in soup (though it is probably essential that you be very hungry first).

The name fireweed suggest a negative connotation— a weed is, after all, something humans usually dislike. If we change the name to "firehealer," might we gain an appreciation for the role it plays in reforming ash into soil, colonizing areas that most other species disdain?

Bog Plants

Sphagnum Moss

Labrador Tea

Cottongrass

Bog Laurel

Bog Rosemary

Leatherleaf

Sundew

Small Cranberry

Pitcher Plant

Bladderworts

Sphagnum Moss

Sphagnum spp.

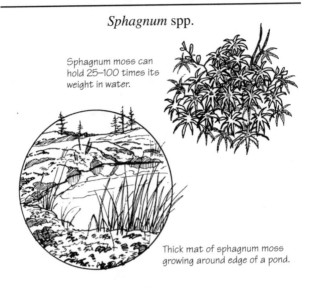

Sphagnum moss can hold 25–100 times its weight in water.

Thick mat of sphagnum moss growing around edge of a pond.

Mosses, by definition, are any small green plant having leafy stems but no flower, and growing so close together as to form velvety cushions. Mosses lack a vascular system (the rigid stem for carrying water and minerals up from the roots), and so they always grow close to the ground. A slender bristlelike stalk, or seta, bears a spore case at its end, the result of a rather complicated sexual reproduction process. Moist habitats are a general requirement for mosses; look for them on rotting logs, on the forest floor, and around wet areas. However, some species even colonize sand dunes.

Nearly twenty species of sphagnum moss are found in the Great Lakes region. If you're standing in a bog, look down at the spongy, hummocky ooze around your ankles. You're probably standing on sphagnum moss.

Bogs make up a significant portion of the North Woods landscape, and they offer a unique community of plants. Bogs developed after the retreat of the last glacier, when numerous depressions left on the landscape filled with water and became lakes. In those lakes that had no inlet or outlet, organic matter built up from the annual deposit of leaves around the lake, and very slowly a mat of vegetation began to crawl out from the shoreline. Sedges in particular invaded the edges, forming a shelf of plants out over the water and beginning the process of building the bog.

Other plants followed, among them sphagnum moss, capable of soaking up as much as twenty-five to one hundred times its weight in water. Sphagnum eventually became the dominant plant in the bog, serving as the spongy carpeting on which nearly every other plant grew. That sponge was not only cold and wet, but highly acidic due to the oxygen-deprived, stagnant lake water on which it grew. Sphagnum further acidified the bog by releasing hydrogen ions into the water in exchange for minerals and by extracting sulfur from the atmosphere and releasing it as sulfuric acid. The acid and cold conditions make a hostile medium for the decomposition of bacteria and other organisms, so, the bog mat built upon itself yearly as plants died and simply remained, eventually creating deep layers of peat.

The processes continue today. The northern bog provides one of the least hospitable habitats for plants. Those few plants that can tolerate the constant wet, cold, acidic conditions also must contend with a lack of nutrients because so little decomposition takes place. Every plant in the bog employs a set of adaptations to survive these hardships. Four species of carnivorous plants have resolved the lack of nitrogen by trapping insects for their dietary needs.

Other bog dwellers have made desertlike adaptations: inrolled leaves, spongy-succulent tissues, waxy, hairy, or leathery leaf surfaces—all characteristics found in desert plants that must conserve water. But why would plants standing in water need to conserve it? Probably for three reasons: (1) the water in a bog is so acidic and cold that plant tissues can't readily absorb it; (2) nutrients, as well as water, can be lost through transpiration, and in bogs, nutrients are in such short supply that the plants must retain all that they can; (3) the moss remains frozen often into the early summer, yet sunny, warm conditions prevail, so the sphagnum must conserve water until thawing is complete.

Bogs often are referred to as "quaking." A bog "quakes" when the sphagnum mat has covered the surface of the water but hasn't completely filled in the bottom with organic debris. The bog mat literally floats on the water, and rolls and "quakes" when one walks on it.

Due to the exceptional preservative qualities of sphagnum peat, in the last two centuries seven hundred human corpses have been recovered from northern European bogs. Some bodies date back to the Bronze Age, about three thousand years ago, yet remain in remarkably good condition. Their skin shows little drying, although it's usually dyed dark brown due to the humic acid in the peat.

Of equal scientific interest are the pollen grains found perfectly preserved in the layers of peat. From study of the pollen, scientists can determine the historic succession of vegetation in any given area. From that information, they theorize about climatic conditions and the nature of historic ecosystems.

Sphagnum moss has had a host of historical uses. In World War 1, it was used in place of surgical cotton. It absorbs liquids three times as fast as cotton, in amounts three times as great; it retains liquids better, reducing the need for dressing changes; it's slightly antiseptic due to its acidity; and it's cooler, softer, and less irritating. Its absorbent quality made it useful as a "diaper" for babies of various Native American tribes.

Today sphagnum moss makes a superior mulch, a packing material for shipping plants, and insulation. Commercial peat mining is done in Minnesota, Michigan, and Wisconsin, primarily for horticultural purposes.

A tealike preparation was once recommended for eye diseases and as a cure for hemorrhaging. Burning moss may be used to create a smudge screen of smoke that protects crops from frost. In Great Britain, compressed peat is cut into blocks, dried, and used as fuel, though it burns cool and smoky compared to wood.

Humans make very little use of mosses as food— Laplanders used sphagnum moss in bread, but it was described in an ancient botany book as "wretched food in barbarous countries."

ACTIVITY

Pull on your rubber boots or your old tennis shoes and walk onto the edge of a bog. Sphagnum moss grows only on the surface and dies below water. Hence it floats on the surface without putting down roots. Grab a handful and see how easy it is to pull it up. Squeeze the mass and a stream of water should pour out.

Sphagnum Moss

Sphagnum moss doesn't tolerate much abuse—walking in bogs will rapidly create trails—so if you do much bog exploration, alter your route frequently, and pick your way carefully to avoid harming the many rare orchids and insectivorous plants common to bogs.

Labrador Tea

Ledum groenlandicum

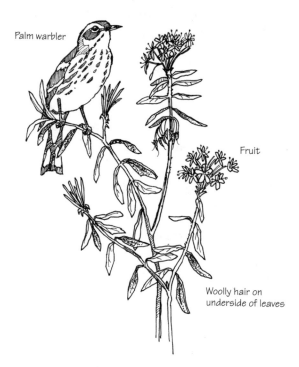

Palm warbler

Fruit

Woolly hair on underside of leaves

This evergreen shrub grows from 1 to 3 feet tall, with an upright umbrellalike terminal cluster of white flowers. Each blossom blooms in late May or early June. To prevent water and mineral loss, the untoothed leaves are thick, succulent, and rolled inward. Woolly hair covers the underside of the leaf, rusty brown in color if it's an older leaf, white if it's young. Dense hair insulates the twigs as well, and the crushed leaves release a pleasant fragrance that is an identifying trait.

Labrador tea prefers the drier edges of bogs or the high side of hummocks within the bog. In a dry early summer a bog may put on quite a show with the simultaneous flowering of Labrador tea and bog laurel.

Labrador tea is also called muskeg tea due to its habitat preference. The common name, Labrador tea, was derived from its historical use by people in Labrador.

As a source of food for wildlife, Labrador tea is of minor browsing value to white-tailed deer and moose. Carl Linnaeus, the Swedish naturalist, reported that farmers mixed the leaves with corn to deter mice. The tannin-rich leaves were historically used in Russia for tanning leather.

ACTIVITY

To make a mild tea, just as colonists did during the Revolutionary War, steep the dried leaves for 5 to 10 minutes.

Cottongrass

Eriophorum spp.

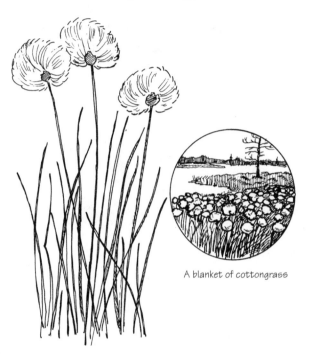

A blanket of cottongrass

Cottongrass looks just like its name indicates it should—the flower resembles a tuft of cotton at the tip of a grasslike long stem, often 3 feet or taller. Depending on the species (there are six species in Wisconsin, Michigan, and Minnesota), the flowers may grow in a dense cluster or clump resembling a pincushion, or they may grow as scattered solitary individuals. During June in some bogs like the Powell Marsh in northern Wisconsin, cottongrass is so thick that the bog appears to be covered with snow—a snow that waves in the wind!

Cottongrass

Cottongrass belongs to the sedge family, a family whose members only a botanist can differentiate. The general family characteristics of the sedge family include: three-ranked leaves (they spiral up the stem in three different locations before repeating the pattern) and three-sided stems, triangular in cross section. The helpful ditty goes, "Sedges have edges."

Bog Laurel

Kalmia polifolia

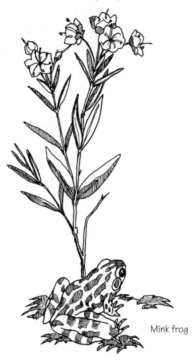

Mink frog

A low evergreen shrub, bog laurel, also called pale laurel, has long, narrow, untoothed leaves that are rolled inward, apparently for water and mineral conservation. The opposite leaves, white underneath, provide the key identifying factor. The leaves are somewhat similar to those of bog rosemary, but it has alternate leaves. Here's my simple mnemonic: Laurel and Hardy had opposite personalities, and this hardy laurel has opposite leaves.

The exquisite five-lobed pink to rose flowers rise above the leaves on thin stems that usually hang down.

Flowering in late May to early June, the saucer-shaped blossoms look like small smiling faces. I can think of no better reason to risk getting your feet wet than to get a closer look at these beautiful flowers.

The genus name *Kalmia* honors the eighteenth century Swedish botanist Peter Kalm.

Bog Rosemary

Andromeda glaucophylla

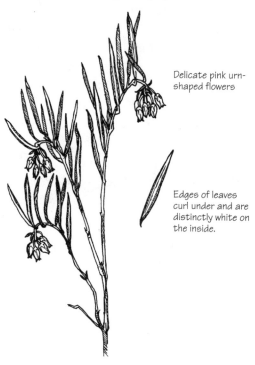

Delicate pink urn-shaped flowers

Edges of leaves curl under and are distinctly white on the inside.

og rosemary is another low, evergreen bog shrub with long, narrow, alternate leaves that are inrolled to conserve water and/or minerals. The underside of the leaf has tiny bluish white hairs, and in May and June small white to pink, urn-shaped flowers grow at the branch tips in a small cluster. Red-brown seed capsules form in the fall and split, releasing the tiny airborne seeds. Bog rosemary requires two years to grow from bud to flower to seed to new plant.

Bog rosemary spreads vegetatively throughout the bog by horizontal creeping rootstocks. The leaves have little wildlife or domestic animal value because they contain the poison andromedotoxin.

The Swedish botanist Carl Linnaeus, while on a bog hike in 1732, chose the genus name *Andromeda* for bog rosemary. He wrote, "This plant is always fixed on some little turfy hillock in the midst of the swamps, as Andromeda herself was chained to a rock in the sea, which bathed her feet, as the fresh water does the roots of the plant. Dragons and venomous serpents surrounded her, as toads and other reptiles frequent the abode of her vegetable prototype."

The species name *glaucophylla* derives from the botanical term "glaucous," which describes the fine bluish or whitish bloom on the leaves, and *phyllo,* meaning "leaf."

Leatherleaf

Chamaedaphne calyculata

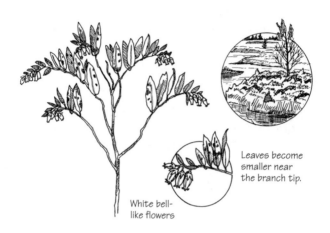

Leaves become smaller near the branch tip.

White bell-like flowers

A tall (1 to 4 feet), evergreen shrub, leatherleaf often dominates in a bog. The untoothed, alternate thick leaves are leathery, hence the name, and are often covered with rusty scales. But the distinctive identifying trait is how the leaves clearly decrease in size near the tip of the plant. Leatherleaf's white late-April to early-May flowers hang bell-like in a row from one side of the upper branches; they are similar in shape to the more often noticed flowers of blueberries.

Leatherleaf grows in dense thickets on top of the bog's sphagnum moss ground layer, spreading by root suckers. By doing so, it assists the successional process that occurs, in which upland plants gradually invade the drier areas of bogs.

Leatherleaf is well adapted to heavy snows. When it has been completely covered by snow, nearly all its flowers will bloom, but if it has been exposed to winter winds due to little snow, few if any flowers will bloom.

In winter, leatherleaf turns a russet color, and since it often dominates the bog shrub layer, many acres may appear a deep rusty brown. Historically fire tended to reduce leatherleaf's dominance in bogs, but with the increased control of fire, leatherleaf has prospered at the expense of other species like cranberry.

Chamai means "on the ground," and *daphne* means "laurel," a reference to the plant being a low evergreen. *Calyculata,* from *calyx,* Latin for "cup," means "shaped like a cup," an obvious reference to leatherleaf's cup-shaped flowers.

The ripened fruits form upturned capsules that hold the seeds until the wind shakes them out. The fruits and buds provide winter food for sharptail grouse. White-tailed deer, cottontail rabbits, and snowshoe hares browse the leaves.

ACTIVITY
Feel the leathery texture of the leaves. That tough outer surface serves to hold in water and minerals in the desertlike environment of the bog.

Sundew

Drosera rotundifolid

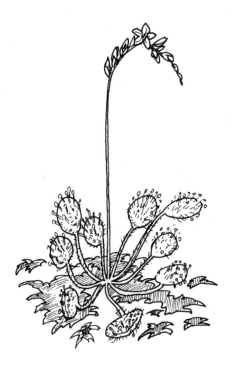

A tiny resident of bogs and cedar swamps, sundew produces leaf stems that grow in a circular rosette a little smaller than a half dollar. Each leaf blade expands at its tip into an oval "sun" that is covered with thick hairs. The hairs project like the rays of the sun, each with a glistening drop of sticky fluid at its end, like dew. The minute flowers bloom white or pink in July, forming all on one side of a long (4 to 9 inches) slender stalk.

The carnivorous sundew attracts diminutive insects like mosquitoes with its attractive rosy coloration, sweet smell, and sparkling fluid-tipped tentacles, trapping them in the viscous droplets. The leaf gradually (10-15 minutes) folds over the captive, and then secretes an enzyme to begin digesting the insect. Complete digestion takes several weeks. While not as dramatic as a Venus's-flytrap, which closes instantaneously, the sundew can take larger prey—one botanist in Michigan watched a monarch butterfly fail to free itself from the syrupy "dew."

Sundews had a surprising number of historical uses as a red ink, a wart and corn remover, an antispasmodic, and a natural flypaper to be hung indoors.

The Greek word *Drosera* is derived from the word for "dewy," and *rotoundifolia* defines itself, "round leaf."

Activity

Darwin experimented at length with the insectivorous plants, in one set of experiments attempting to feed inedible substances to the sundew. The sundew, after a brief embrace, would drop the unwanted substance and wait for its next opportunity. Repeat Darwin's experiment using live insects as well as other materials, and note the response of the plant.

Small Cranberry

Vaccinium oxycoccus

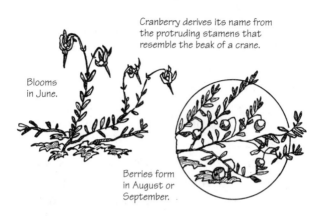

Cranberry derives its name from the protruding stamens that resemble the beak of a crane.

Blooms in June.

Berries form in August or September.

Cranberry is a small, slender, creeping shrub with slender, woody stems. It is hard to spot unless you are on your knees in the bog. The alternate, simple, smooth-edged leaves are tiny, less than 3/8 inch long, and are white underneath. The leaf edges roll in, the leaf tips are pointed, and the entire plant is hairless. Cranberry blooms pink or rose in small clusters on slender stems, usually in June. The petals of the flowers peel back and the stamens extend outward. The berries, which ripen in August and September, are red and often speckled, and twice the length of the leaves—a surprisingly large fruit from such a tiny plant.

Cranberry spreads itself over the bog mat by sending out long stems that root at the joints and begin another clonal plant. The leaves, like most bog plants', are evergreen, an adaptation that saves the energy of spring leaf formation.

The berries taste sour, but they make fine sauces and jams. Wildlife seems to ignore the berries, so the fruits may last throughout the winter, though chipmunks, ruffed and sharptail grouse, and mourning doves sometimes eat a few.

The large cranberry (*Vaccinium macrocarpon*) constitutes the commercial cranberry species, and growers have planted large diked acreages in northern Wisconsin. The dikes draw in sandhill cranes, bitterns, and herons for feeding, and in the fall when the berries are flooded for picking, snowy owls have been seen hunting small mammals.

The berries keep well. Early settlers picked their share, and Native Americans used them in trade. An 18th century essay claimed that they were excellent for treating scurvy. The Ojibwa made a tea of the leaves to cure nausea.

Cranberries got their name from the Old German "Kranbere," or crane-berry, in reference to the long protruding stamens that resemble the beak of a crane. *Oxycoccus* comes from *oxy,* meaning "sharp, keen, or acid," and *coccus,* meaning "berry."

Activity

In September or October, put on your bog shoes and explore a bog for cranberries. The berries remain firm all autumn, and will stay in good shape in your refrigerator for months if washed and packed in a sterilized jar. Use the wild berries just as you would the commercial ones. Cranberries contain their own pectin, so they make a good jelly or jam.

Pitcher Plant

Sarracenia purpurea

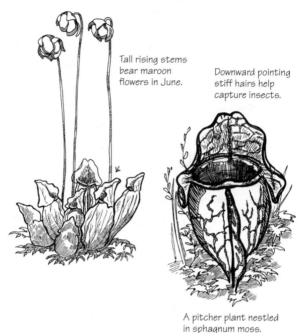

Tall rising stems bear maroon flowers in June.

Downward pointing stiff hairs help capture insects.

A pitcher plant nestled in sphagnum moss.

The pitcher plant is a unique and unmistakable species, possessing a series of large pitcher-shaped or trumpet-shaped, inflated leaves pointing upward from the ground. These leaves are purple or green-veined and mottled red; some suggest they imitate flesh to attract insects. The tall maroon flowers, rising about 2 feet on a single stem, appear in June and July. The blossoms nod upside down like umbrellas. Pitcher plants are the official flower of Newfoundland.

Pitcher plants are one of the four carnivorous plants of the North Woods. While certainly not numerous or common, they are highly visible to people willing to walk out onto a sphagnum bog or into a wet grassy meadow. Pitcher plants have solved the problem of nutrient scarcity in bogs by using fragrance and color to lure insects into their red and green "pitchers." Downward-pointing stiff hairs help prevent insects from crawling back out once they've entered, and a wetting agent reduces the surface tension of the solution in the pitcher, making it difficult for insects to remain on top of the water. Once trapped, the insects decompose through bacterial action in the water.

Even frog skeletons have been found in pitcher plants, apparently victims of their own desire to steal flies from the vessel. Spiders sometimes spin webs in the vessel to take advantage of their attractiveness to insects.

Some insects spend their entire lives in the rainwater of the pitcher plant, apparently immune to the digestive juices. They survive by eating the other insects that drop in. (This process of one species benefiting while the other is unaffected is called commensalism.)

Note: One dissenting botanical source claims that the decaying insects provide food not for the plant, but for the larvae of the insects that cross-pollinate the flowers.

Observation

Look into one of the vessels to detect what the dinner menu is for that day. Watch to see if you can observe an insect crawling down the pitcher and falling in.

Bladderworts

Utricularia spp.

Flat-leaved
bladderwort

These aquatic plants, with long, fine stems (up to 3 feet) and threadlike leaves, grow on muddy shorelines or float under the water surface and are seldom seen except when flowering. The lipped, usually yellow blossoms (some species are purple-flowered) rise above the water surface in July on a naked stem and look a bit like small snapdragons. Attached to the filamentous stems below the water are tiny bladders only millimeters in size and best seen using a hand lens or microscope, which often contain microscopic insects trapped within.

Bladderworts may be found suspended, or sometimes anchored, in the open-water pools of bogs and other shallow stagnant waters. Bladderworts capture minute insects in a hollow bladderlike trap enclosing a partial vacuum that is triggered by hairs near its opening. The bladders work a bit like squeezing all the air out of a ball and then letting go, resulting in a sucking inrush of air. Insects swim by, brushing against the bladderwort's hairs; the compressed bladder releases and sucks in the insect; and an elastic trapdoor mechanism snaps shut, preventing escape. The whole process takes 1/460 of a second, making bladderwort the fastest-moving plant in the world. The insect, however, must decay for several weeks before its fluids can be absorbed. Neltje Blanchan, in her 1901 book *Nature's Garden*, wrote that a sign should be posted above the bladderworts reading "Abandon hope, all ye who enter here."

Bladderworts provide food for a number of large mammals that feed in wetlands, such as moose and deer, as well as some waterfowl.

Utricularia, is from Latin for "little bladder."

OBSERVATION

Examine a tiny bladder with a hand lens to look for insects trapped inside. Smell the blossom. Horned bladderwort exudes a wonderful but difficult to describe scent, unlike any other flower. John Burroughs wrote that it is "perhaps the most fragrant flower . . . . Its perfume is sweet and spicy in an eminent degree."

Wetland/Open Water Plants

Water Lilies

Common Cattail

Purple Loosestrife

Pickerelweed

Blue Flag Iris

Horsetails

Arrowhead

Wild Calla

Water Lilies

Yellow Pond Lily *Nuphar advena*
White Water Lily *Nymphaea odorata* or *tuberosa*

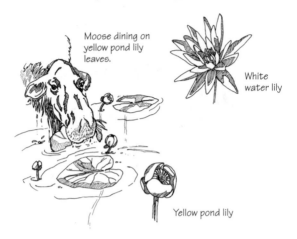

Moose dining on yellow pond lily leaves.

White water lily

Yellow pond lily

Yellow pond lily has yellow flattened globelike flowers with six sepals that curve inward to form a cup. The petals are less than 1/2 inch across and are rather inconspicuous inside the globe of sepals. White water lily has many large white petals, with a center full of long yellow stamens. Two species of white water lily may be found, one which is wonderfully fragrant (odorata), the other without fragrance (tuberosa).

Both lilies float on the surface of quiet, shallow water. And both have similar, very large (9 to 12 inches), oval, shiny green leaves that are deeply notched at their base. However, white water lilies have a right angle formed in their notch, while yellow water lilies curve gracefully at the notch. The spotless white blooms among platters of shiny green leaves, united with a fragrance, reminiscent of fresh oranges, elevate the white water lily to other-worldly status.

The ephemeral flower of the white water lily opens from morning to midafternoon for a period of three to five days until it is pollinated by the myriad of insects drawn to the showy bloom. The sepals then close the flower, and it is drawn underwater as the stem begins to coil and shorten. The flower head gradually changes into a seed vessel, which breaks off and floats to the surface, where it may follow any current to a new bay before sinking and germinating.

Yellow pond lily, whose flowers are neither flamboyant nor fragrant, bears the common names of bullhead lily, spatterdock, and cow lily.

The leaf stems carry oxygen through several hollow tubes to the starchy roots anchored in the mud. If you pick the underwater stem, you can blow through it like a straw. The leaves are waterproof and provide the perfect lawn chair for frogs.

Moose prize the leaves; beaver, muskrat, and porcupine eat the entire plant; and the seeds are a moderately useful duck food.

Yellow pond lily tubers were reportedly harvested for food by Indian women, who either used their feet to expose and break off the rhizomes under the water or simply stole the tubers from muskrat houses.

OBSERVATION

If you see either lily floating in a marsh, note its enormous rootstock. The roots remind me of a knobby, fat softball bat, but larger yet.

Common Cattail

Typha latifolia

Cattails provide excellent nesting and cover habitat for birds and ducks.

Cattail leaves reach 6 feet or higher, and are long and narrow, much like green-veined ribbons. The flower stalk may extend higher yet, to 8 feet, and ends in two flowering spikes. The upper spike is the male flower head, composed of the stamens. These shed their pollen onto the lower female flower structure, which looks like a brown corndog. The tubular stem is filled with large air cells.

The narrow-leaved cattail has a space between the male and female flower heads; the common cattail has no space. A hybrid can be distinguished only by an magnified examination of the flowers.

The seedheads contain hundreds of thousands of tiny wind-borne seeds. The seeds germinate on open mud flats and grow at an astonishing rate, soon sending out horizontal rhizomes from their roots. In the first year, one seed may produce a rhizome system many feet long, with one hundred shoots. These stems push up all around the parent plant, forming a colony of clones.

Cattails are so productive that their dry weight per acre can reach twenty tons (compared to wheat at three tons). This organic matter, laid down on a marsh, eventually builds, and cattails push out farther into the marsh, thus slowly reducing the size of a wetland.

Since cattails grow so rapidly, some waterfront owners have lost open water areas to cattail growth. The best solution is to ensure that muskrats are content in that habitat. Muskrats eat the rhizomes, leaves, and the base of the cattail, as well as using the stalks for cover and for building their rounded huts. Muskrats generally keep a balance between open water and cattails that makes both wildlife and people happy.

Cattails provide excellent nesting and cover habitat for a host of animals, including red-wing and yellow-headed blackbirds, marsh wrens, ducks, bitterns, and rails. Geese eat the starchy underground stems.

The fluffy seedheads have been used for stuffing pillows and blankets. The Ojibwa wove cattails into mats for the sides of their wigwams.

ACTIVITY

Pull a young stem from the water, and remove the spongy sheath around the inner stalk. The first 4 inches or so tastes like a young cucumber.

Purple Loosestrife

Lythrum salicaria

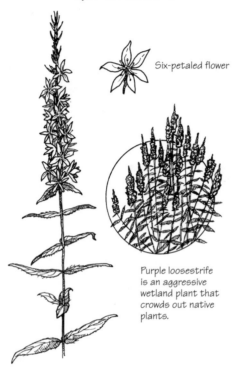

Six-petaled flower

Purple loosestrife is an aggressive wetland plant that crowds out native plants.

Purple loosestrife's 2- to 5-foot-high, tapering spikes of magenta flowers often dominate wet meadows and marshes. The flowers have six petals, unlike fireweed, a plant it is sometimes confused with, which has four petals. The downy, stemless leaves grow opposite in pairs or threes.

Purple loosestrife blooms in midsummer and may completely dominate some wetlands such as over 1,000 acres of a wildlife refuge in New York state. The imported purple loosestrife threatens native vegetation because it aggressively crowds out cattails, sedges, and other native wetland plants. Without these native plants for food, nesting, and shelter, wildlife populations struggle to hang on to traditional community sites. Unfortunately, very few species use the seeds and roots of purple loosestrife, though many insects and butterflies enjoy the blooms.

Purple loosestrife is successful largely because it can sprout from seed in areas already densely populated by other plants. And it produces seeds in abundance—one plant may shed 90,000 tiny wind-borne seeds in a year. Once established, the taproot sends up new stems, producing a thick clump of loosestrife. Wisconsin, first invaded in the 1920s by this European immigrant, is the most seriously infested in the southeast, but there are areas in the north where loosestrife abounds, and it continues to spread. The diversity of many marshes and meadows could soon become a monotype of royal purple.

ACTIVITY

You can stop purple loosestrife by hand pulling before they set seed, if the population is small or just getting started. Check a plant identification book or get the color flyer produced by the state department of natural resources. If you see any loosestrife, pull the plants up by the roots—they can reproduce if just cut off. Show no mercy; your wetland flora and fauna will thank you.

Pickerelweed

Pontederia cordata

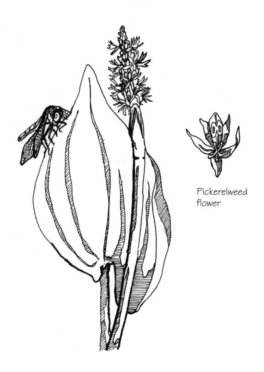

Pickerelweed
flower

The tall blue-purple flowers rise on a 1- to 4-foot spike above large, glossy, heart-shaped leaves with rounded lobes at the base. *Cordata*, the Latin species name, means heart. The leaves may grow so thick along a stream edge or shallow slough that they form an impenetrable barrier for canoeists and anglers. The blossom fades after just a day and then develops into a single seed. Some people find the fragrance unpleasant. Flowering in July, pickerelweed is often found alongside the more fragile white flowers of arrowhead, painting a shoreline in rich colors.

Pickerelweed rapidly colonizes sloughs by sending underground rhizomes that poke up through the bottom mud as new clones of the parent plant. As with most emergent plants, the stems are laced with air compartments to provide architectural support and to pump air to the roots.

The shallow waters where pickerelweed abounds provide excellent habitat for pike to breed and lay eggs. This led to a belief that pickerelweed could spontaneously generate pike and pickerel, hence the name. In 1558 Gesner advanced the theory that the pickerel plant along with other "glutinous matter" would, with the help of the sun's heat in certain months of the year, transform into pike. This strange idea about the generation of living creatures by other sixteenth century botanists, who thought that sheep were spontaneously generated from the blossoms of certain trees.

Each flower produces a bladderlike fruit containing one seed, which is sometimes eaten by wood ducks, black ducks, and muskrats.

OBSERVATION

Look closely at the small purple flowers to see the bright yellow dots that are often arranged in a "butterfly" imprint on the upper petals. Damselflies, dragonflies, and bees spend a good deal of time among the blossoms and leaves.

Blue Flag Iris

Iris versicolor

B lue flag's long leaves (up to 3 feet) are like pointed swords. Each leaf has parallel veins, but is slightly curved and shorter than the flower stem. The showy violet-blue flowers spread out nearly flat, the longer sepals appearing like extended wings, with a greenish yellow blotch at the base and beautiful white veining throughout. A thick fleshy horizontal rhizome extends underground, distributing blue flag in clonal patches, though for some reason it seldom develops into large colonies in its favored habitats—sedge meadows, marshes, and stream banks.

Iris means "rainbow" in Greek, indicating the variety of colors represented in the genus. Henry David Thoreau thought the blue flag was "a little too showy and gawdy, like some women's bonnets."

The offensive tasting but odorless roots contain irisin, which causes indigestion in humans and cattle. Native Americans in New York in the 1700s did use the boiled root as a remedy for leg sores, and the Ottasses Indians in Georgia used a root decoction as a cathartic whenever they fasted for seven days to rid their village of sickness. As food or cover for wildlife, the blue flag fails to provide much value compared to other wetland plants.

OBSERVATION

Sit down close to a blue flag iris and watch the pollination of its blossom. The unique shape of the blue flag flower ensures cross-pollination. Bees are strongly attracted to violet and to large showy flowers, both characteristics of the blue flag. When a bee lands on one of the sepals, it follows the yellow lines (nectar guides) into the opening under the female part of the flower, scraping off any pollen on its back from a previous visit to another blue flag. As it crawls further in, the bee gets new pollen from the male part of plant, and then finally reaches the base of the flower, where it feeds on the nectar. After drinking its fill, the bee leaves through an opening and is off to another blue flag.

Horsetails

Equisetum spp.

Horsetails are grouped with the "fern allies," those plants that are similar reproductively and structurally to ferns, but that have needlelike or scalelike leaves. (True ferns have large, flat fronds with branching veins.) The hollow, fluted, jointed stems offer the easiest identifying trait; you can pull the stems of many species apart and put them back together at the joints. The stems are bamboolike, and the joints are ringed by a collar of black-and-white tiny, toothlike leaves. Some species, like common horsetail (*Equisetum arvense*), have long, needlelike branches whorled around the base of the joints, while other species, like scouring rush (*Equisetum hiemale*), are without branches. Only twenty-five species of horsetails are known throughout the world, ten of which occur in the North Woods.

Equisetum means horse bristle, from the taillike appearance of the many-branched species. Common names include scrubgrass, shavegrass, polishing rush, and gun bright.

Horsetails grow from underground, jointed rhizomes, which send up shoots at each joint. They also reproduce sexually. Usually the fertile stems arise first in spring, pink and topped by cone-shaped sporangia. They release their spores, die back, and then the green sterile stems emerge. Horsetails spores only live a few days and must begin to grow rapidly, or they will die. Both the stems and the branches are photosynthetic.

Horsetails haven't changed very much since the Coal Age, 300 million years ago (though they're certainly much smaller than the treelike plants they were back then), and they can be thought of as living fossils.

Horsetails grow in diverse habitats, from wet soils along rivers and in marshes (where the bushy underground root and rhizome growth acts as an anchor for the soil) to well-drained sites like highway and railroad embankments.

The grooved stems of horsetails, particularly scouring rush, contain deposits of silica sand. Colonists used them to scrub and scour pots and pans. (If the plant is burned, the ashes may also be used in a scouring mixture, because the silica remains.) Some tribes even used the abrasive stems as a natural sandpaper to polish wood, file fingernails, sharpen arrows, and so on. The stalks of scouring rush were boiled and the liquid used as a hair rinse to eliminate fleas, lice, and mites. The heads of the reproductive shoots were eaten to cure diarrhea.

Field horsetail contains equisetic acid, which is supposedly a potent heart and nerve sedative but which is poisonous when taken in high doses. Other genus members contain equisetine, an alkaloid that poisons livestock.

Whistling swans and snow geese eat the stems and rootstocks; moose, and to a small degree black bear and muskrat, eat the whole plant.

ACTIVITY

Pick a stem of horsetail from a colonial patch and pull it apart at the joints. The stems of most species should fit right back into the joints from which they were separated.

Arrowhead

Sagittaria spp.

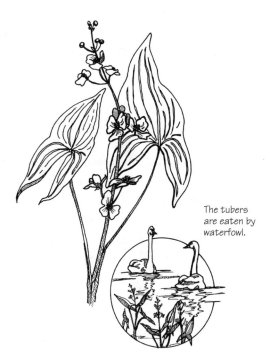

The tubers are eaten by waterfowl.

A rrowhead has, unsurprisingly, arrowhead-shaped leaves with two long, pointed lobes. The flowers appear in July in delicate whorls of three with three green sepals, three snowy white petals, and a golden center on a 1- to 3-foot tall, naked succulent stem. Long, narrow leaves like grass blades ripple underwater, their slim width an adaptation for withstanding the power of underwater currents.

Arrowhead belongs to a group of marsh plants called emergents, meaning they are rooted in shallow water bottoms, but grow well above the surface of the water. All emergent plants send out clonal rhizomes under the water bottom to form dense colonies that help protect individuals from the hazards of winds and waves. The many stems also slow down water flowing by, allowing nutrient-rich sediments to collect around the plants. Arrowhead leaves can be quite broad or like long, narrow lances, perhaps due to fluctuating water levels. Seeds colonize exposed soils during low water levels, and once the plants are established, asexual (vegetative) reproduction dominates.

The tubers have significant wildlife value for wood ducks, canvasbacks, geese, swans, and muskrats. Native Americans and settlers collected the tubers and boiled them like potatoes, giving rise to other common names such as duck potato and wapato. Lewis and Clark ate the tubers in large numbers while spending a winter at the mouth of the Columbia River and reported that Indian tribes in Oregon used the tubers as a principal article of trade. When fresh, the tubers contain a somewhat bitter, milky juice. A medicine for indigestion and a poultice for wounds were also made from the tubers.

Sagittaria is derived from the Latin *sagitta*, for "arrow."

Wild Calla

Calla palustris

Wild calla
grows in bogs.

Wild calla grows low to the ground, with smooth, long-stemmed, parallel-veined leaves. The unique brilliant white "petal" (really a spathe) acts as a silky hood for the real flower, a golden club of clustered tiny blossoms. The glossy heart-shaped leaves arise from long creeping stems, which can float and which help calla to form large clonal colonies.

Wild calla blooms during June in bogs and shallow water margins. Wild calla resembles the cultivated calla lily, but they're not in the same genus.

The fruit forms a clustered head of brilliant red berries. Native Americans used the tuber for food and as a poultice. The seeds and rootstocks, when dried, can be ground into an unpalatable but nutritious flour, but eaten raw the tubers are acrid, causing an intense burning sensation in the mouth. One current source (Lyons and Jordan) claims that the rootstock must be "roasted for several hours, and then dried for six months before being ground into flour." Better to enjoy the brief beauty of the flower than destroy it for such a questionable culinary return.

Palustris is Latin for "swamp," and while the origin of *Calla* is uncertain, it is believed to come from the Greek *kalos*, meaning "beautiful." Our youngest daughter Callie likes to think this interpretation is true.

Also called water arum, swamp robin, and female or water dragon, the often pure stands of wild calla grace a dark-soiled wetland with a candle flame of white.

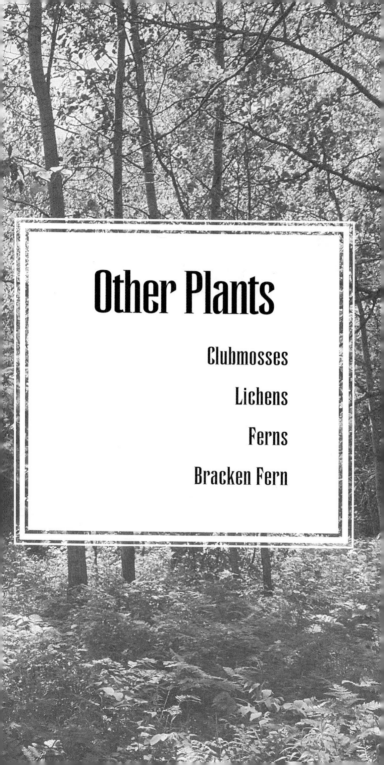

Other Plants

Clubmosses

Lichens

Ferns

Bracken Fern

Clubmosses

Lycopodium spp.

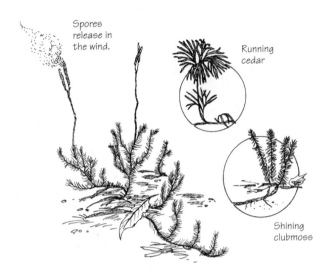

Spores release in the wind.

Running cedar

Shining clubmoss

Runner of "wolf-claw" clubmoss

Fern allies include clubmosses, horsetails, spikemosses, quillworts, and whiskbroom ferns. Fern allies are a group of fern relatives that have the same structural characteristics as ferns—a vascular system for support and "plumbing," and an alternating-generation life history that progresses from spores to a sexually reproducing form and back again. The difference between ferns and their allies is that ferns have large flat fronds with branching veins, while the allies have needlelike or scalelike leaves with just one unbranched vein. They don't look a bit like ferns or much like each other. Clubmosses look a bit like mosses but are really more like miniature conifer trees. Horsetails look like a piece of jointed bamboo, often with coarse hairs. Quillworts might be mistaken for clumps of grass or the long stems of young onions.

Clubmosses, like the "princess pine" and the "running cedar," are the fern allies most often noticed by the average hiker. Christmas-wreath makers pick them extensively. All of the six species found in the North Woods superficially resemble a conifer seedling, but clubmosses seldom grow taller than 6 inches and they produce spores, not cones with seeds.

Clubmosses are perennial evergreens. They vegetatively reproduce by sending out runners along or just below the ground surface, which root down at intervals. New shoots may pop up at the roots, or the stem may continue farther, walking itself slowly through the woods while the tail end dies off.

In late summer and early fall, little club-shaped structures grow from the tips of the shoots. Some of the clubs are on slender stalks that look like a candelabra set for a romantic vole's dinner. Each club packs thousands of spores. If you tap one of them when they're ripe, the yellow spores billow out, catching the wind and floating to their new homestead site in the woods.

Clubmoss spores found their way into a remarkable number of early products, such as coatings for pills, powders for irritated skin, and mordants for woolen dyes. Due to their consistent size, the spores were used as standards for microscopic measuring. Some amazingly creative person even discovered the spores to be explosive, and they were used to produce the pan flash for old-time cameras.

The complex reproductive cycle of clubmoss closely resembles that of ferns. It takes clubmoss spores approximately 17 years to grow through their alternate life stage and into the next-generation spore-bearing plant. Laws to protect these overlooked and underappreciated plants are overdue, particularly given that some clubmoss plants have lived up to 1200 years!

OBSERVATION

Follow the runner of a clubmoss from its current growth end back to its origin, taking note of the many rootings of the runner along the way. A clubmoss may literally walk its way to a new location where conditions are right, and take root there, sort of a squatter's rights approach to reproduction.

Lichens

Genus and species name vary

Lichens take a variety of forms, ranging from leafy structures to papery flakes.

Lichens, plants without distinct stems, leaves, or roots, most commonly colonize rock surfaces, tree branches and bark, and forest soils. They can, however, be found virtually anywhere in the world and on nearly any surface.

Botanists remain unsure how to classify the lichens because they are a union of a fungus and an alga in one plant. The nature of the union is also unclear. The alga and fungus may act cooperatively for one another's mutual benefit, the alga performing the food making through photosynthesis, the fungus holding onto needed water and offering protection from injury. Or the fungus may be a parasite of the alga host, deriving unequal benefits from the relationship. The host-parasite theory appears most likely, for if the alga is separated from the fungus, the fungus will die while the alga will live. Only the fungal cells in contact with the algal cells can take in the nutrients from photosynthesis. The fungal parts not in contact with the alga serve as supportive and protective tissue. Look at a lichen under a microscope and you will see a branching network of fungal filaments with green algal cells woven into them.

Lichens pioneer habitats where few other plants could live, carrying life to K2 in the Himalayas at 18,000 feet, to rocky outcrops near the North and South Poles, and to desert rocks too hot to touch. Their ability to colonize bare rock and then hold onto soil particles is essential in creating a window of opportunity for other vegetation to get started. Through the solvent action of acids they secrete, lichens help disintegrate rocks, rapidly decomposing micas and garnets, gradually decomposing calcareous rock, but barely scratching quartz.

Lichens are tough—one lichen colony was exposed to 268 degrees centigrade for seven hours, and the colony still resumed normal metabolism. A few lichen colonies are estimated to be two thousand years old.

Lichens are divided into three types: (1) crustose, meaning crusty or flaky—these often look more like a tint of color on rocks than a plant; (2) foliose, meaning leafy *(folio)* or papery—look for ruffled mats or leaflike structures growing on rocks and trees; and

(3) fruticose, meaning stalked or branching—the most specialized of the lichens, a group including British soldiers, pixie cups, and reindeer lichen.

Lichens sop up water like a sponge through capillary action. Water content varies, from 2 to 10 percent of dry weight on dry days to 300 percent of dry weight on wet days.

Lichens provide an important winter food for Arctic browsers like caribou and domesticated reindeer but have marginal importance for deer, moose, and grouse in the North Woods. Lichens also provide nesting material for birds like wood pewees, hummingbirds, and blue-gray gnatcatchers.

The ancient Egyptians ground lichens for flour, and the Swedes made bread from reindeer lichen, but all lichens taste bitter and are irritating to the intestines. Most often lichens serve as famine food. Sir John Franklin recorded in his 1829 diary that rock tripe saved his party from starvation in the Arctic seas. But his men were also eating leather from all their equipment at the time. The lichens were said to be causing severe illness in the men, but who knows which did the most damage.

Lichen-derived alcohol was distilled and sold in Moscow in 1872. By 1893, Sweden had large brandy manufacturing distilleries using lichens as the starch, but the industry failed in 1896 due to the overharvest of the native lichens.

Applying the doctrine of signatures (see the glossary), people found many imaginative uses for lichens: lung lichen (*Lobaria pulmonaria*), for treating lung disease; a lichen found growing on human skulls, for curing epilepsy; and dog tooth lichen

(*Peligera canina*), a lichen with fruiting discs that with imagination resemble a dogs teeth, for curing rabies. In the first century, people thought some species looked like human skin and used them for curing leprosy.

While you may not wish to learn to differentiate the various species of lichen, taking notice of their existence and appreciating the work they do offers basic ecological understanding. Carl Linneaus reminded us several hundred years ago, "Though hitherto we have considered theirs a trifling place among plants, nevertheless they are of great importance at this first stage in the economy of Nature."

ACTIVITY

Take a mass of dry reindeer lichen (a wiry, white/ gray, elaborately branched "moss") and put it in water. Note how much water it will absorb in ten minutes—usually half its weight.

Lichens won't live and reproduce in polluted air. The sulfur dioxide content in the atmosphere can be estimated by studying lichen populations on trees. In England and Wales, a ten point lichen scale is used for this purpose. If you live in a city, note the lichen populations on trees at various distances from a source of air pollution.

Ferns

Genus and species vary

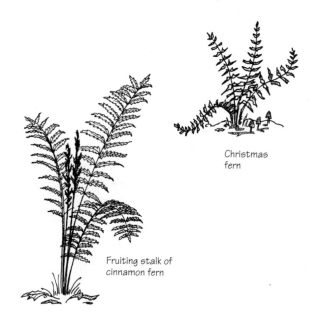

Christmas fern

Fruiting stalk of cinnamon fern

Ferns and fern allies (like clubmosses and horsetails) make up one of the four groups of plants into which botanists have divided the plant world. And though the least populous of the plant groups, comprising just 2 percent of the world's total plant population (twelve thousand species of ferns and one thousand species of fern allies), ferns helped pioneer the lifeless earth some 400 million years ago. In the process ferns locked up solar energy in the form of coal.

There's a lot more to ferns than feathery fronds, though the word "fern" does come from the old Anglo-Saxon *fearn*, meaning "feather," and "pteridophytes." The classification term for ferns and their allies, originates from the Greek word *pteron* for "feather." Just calling a plant a fern, however, doesn't make it a member of the fern family. Asparagus fern may look like a fern but it produces flowers, something ferns don't do.

Ferns lead a secret sex life that only botanists seem to know about. Their public life, that of the fern fronds and the spores they produce, make up only the first stage of their life cycle. The other stage is hidden, usually in a bed of moss. Here a fern conducts a sex life much like any animal might, complete with sex organs, fertilization, and developing young attached to and "fed" by "Mom." Here's how this process works.

Ferns produce spores that are located in fuzzy brown spots or streaks on the underside of the fronds. Each brown spot contains masses of spore cases, and each spore case in turn contains thousands of fern spores. The math adds up to millions of spores produced by a single fern, each spore microscopic and floating on the most fragile of winds to virtually any spot in the hinterlands of the North Woods. Why aren't we knee-deep in ferns? Because spores need to land in a warm, moist spot in order to grow, and the combination of warm and moist isn't always easy to find in the North Country. So most spores die. Those that survive, however, grow into tiny, green, heart-shaped plants about 1/4 inch across that have no true roots, stem, leaves, or conducting system. They look entirely unlike ferns as we ordinarily think of them.

Near the base of this tiny plant, the male sex organs resides. They will release sperm if, and only if, there is water available in which the sperm can swim over to the female organs, which each contain an egg available for fertilization. From this union a "baby" fern develops that is fed by the plant until it grows its own roots, stem, and fronds, and becomes a self-sufficient plant, ready for the next generation of spores.

Ferns also reproduce vegetatively. Most ferns emerge as fiddleheads directly from a thick horizontal stem (rhizome) that grows on or below the soil surface. The plant is very tough, stores food, and may live as long as one hundred years.

Vegetative cloning works well in the short run, as long as the environment doesn't change and require new genetic qualities. Sexual reproduction offers the best long-term reproductive strategy for any organism because new genetic combinations are created with every new fertilization. The combination of both strategies provides an excellent form of species "life insurance."

Europeans in the Middle Ages concluded that because ferns didn't produce visible flowers or seeds like other plants, the flowers and seeds must be invisible. A legend arose that at dusk on St. John's Day, June 24, ferns produced tiny blue flowers that ripened at midnight into fiery, golden seeds that would fall to earth. If people caught these seeds on a white cloth, they would receive power to work wonders. By placing a few of the seeds in their shoes, it was believed, people could gain the power to become invisible, the ability to see into the past or the future, and the insight to find lost objects or hidden treasures. Best of all, fern seeds could bestow eternal youth.

Since most ferns begin as fiddleheads, all curled up like worms, the logical conclusion according to the doctrine of signatures was that ferns could expel worms from the body and could straighten out the bent muscles of arthritic people. Likewise, since maidenhair fern looks like a ring of hair, *The Herbal or General History of Plants* (1633) recommended its use for hair restoration.

Activity
Place a frond with ripe sori (the brown spots or streaks on the underside of the fronds) on a piece of white paper and shine a warm light on it. If you use a microscope you can watch the sori literally pop as the heat dries them and the spore cases burst forth. If you put a container over the paper, a brown dust will accumulate after a few hours—the dust being the microscopic spores looking for a breeze and a wet landing area.

Bracken Fern

Heridium aquilinum

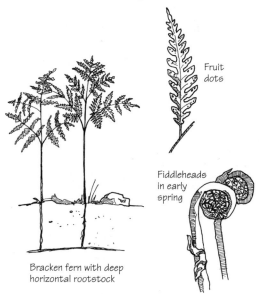

Fruit dots

Fiddleheads in early spring

Bracken fern with deep horizontal rootstock

While the bracken fern looks like it has three large feathery leaves, a single grooved 2- to 2-foot stem produces a single broadly triangular, coarse leaf that is divided into three nearly equal leaflets at the same point on the stem. The leaflets divide into many slightly hairy subleaflets that are narrow and have blunt tips.

In August, fruit dots containing spores, which appear in a line near the edge of the leaflets, mature. Reproduction by spores, however, is rare compared to the vegetative success attained through the rootstocks.

Bracken fern is very common in open woodlands, along roadsides, and in old pastures, outcompeting other ferns in dry, poor, and often acid soils. Bracken competes well in woodlands due to three processes: (1) Horizontal rootstocks the diameter of a pencil, buried 10 inches in the soil, spread from the parent plant, often growing 50 feet in one year, to send up new fronds. The depth of the roots protects the plant from the extremes of weather. (2) The mature leaves are poisonous to animals and are left completely alone by deer and other browsers. Researchers looking for chemicals that will repel herbivores continue to try to isolate the substance that makes bracken so unpalatable. (3) The shade produced by the near continuous canopy of fronds eliminates nearly all competitors. Because of their dominance in many forests, bracken ferns are considered by some to be the "weed" of the fern family.

The young fiddlehead fronds emerge covered with silvery gray hair, and to some they look like an eagle's claw. These fronds may be eaten raw or cooked. Indians used the mature rootstocks in basketry and the fronds as a thatching material.

With virtually no resistance to frost, the visible portion of the bracken fern usually dies in late August. Then its brown, withered fronds allow other species needed sunlight.

Activity

Scrape away the soil around a bracken fern to expose the thick rhizomes that run between virtually every bracken fern in a colony.

Also note the absence of browsing marks on the leaf fronds. Only one or two insects eat the leaves, and their impact is minimal.

Glossary/Sources

Glossary

alternate leaves – leaves that grow singly on a stem, first at one point, and then at another, but never across from one another.

annual – living for only one season and then dying.

anther – the male pollen-bearing part of the stamen.

bract – a modified leaf growing usually at the base of a flower.

catkins – long clusters of tiny flowers without petals, and all of the same sex.

climax stage – the self-maintaining end point of successional change.

clone – a genetically identical individual.

commensalism – a one-sided relationship between two species in which only one side benefits and the other neither benefits nor is harmed.

compound leaf – a leaf divided into leaflets. The leaflets look much like individual leaves, but do not grow from a bud.

disk flower – tiny tubular flowers found in the central part, or "eye," of the flowering head of a composite flower like a daisy.

doctrine of signatures – the belief that a plant's shape indicates its use.

family – a group of related plants. Families are divided into genera (plural of genus), which are further divided down into species.

genus – a group of closely related plants. The genus is the first word in the scientific name of a species, and is always capitalized.

introduced species - a species that is nonnative, or alien, to the area covered in this guide.

leaflet - one of the parts of a compound leaf that looks similar to a separate leaf.

lenticels - breathing pores in the bark of woody stems.

mutualism - a relationship between two species that is mutually beneficial.

opposite - leaves that grow directly across from one another in pairs.

ovary - the female base of a flower that produces seeds.

palmate leaf - a compound leaf with its leaflets radiating out from a central point, like the fingers on your hand or the spokes of a wheel.

perennial - a plant that normally lives for more than two years.

petiole - the stalk that joins the leaf to the plant's stem.

pinnate leaf - a compound leaf with its leaflets arranged like blades (pinnae) of a feather, coming off both sides of a central stem.

pioneer plants – the first plants to invade an area.

pith - the center or core of a twig or stem.

podzolize - to form a podzol soil, one that is very gray in color, with a thin black band of highly decomposed organic material on the surface. True podzols are acidic and nutrient-poor soils.

pollen - the male spores that fertilize the female ovary to eventually form a fruit and seeds.

ray flower - the petallike flowers that encircle the disk flowers in a composite flower like a daisy.

rhizome - underground stems that produce shoots that grow into new, but genetically similar, plants.

samara - a winged seed, like a maple "helicopter."

saprophyte - a plant that gets its food from dead organic matter.

serrate - toothed margin of a leaf.

simple leaf - a single leaf, without leaflets, growing from a bud.

spathe - a large bract that encloses a flower, like a hood.

species - a distinct kind of plant. Every plant is designated by a two-word scientific name. The first word is the genus, the second word is the species.

stamen - the male organ of a flower.

stolons - horizontal stems that creep along the ground surface. Along the way they send down roots at their nodes from which new plants may grow.

succession - the natural and gradual, continuous replacement of one group of species by another.

tension zone - a general area dividing the northern and southern forests, where many northern species reach their southernmost range and many southern species reach their northernmost range.

terminal bud - at the end of the stem or branch.

toothed - having indentations, or teeth, along the edge of a leaf.

tuber - a short, thick underground stem with buds.

vascular plant - plants, like trees and shrubs and wildflowers, that are equipped with tubes for carrying fluids to other parts of the plant. Plants like mosses and lichens are not vascular.

vein - one of a network of tiny channels in a leaf that carry plant fluids.

whorled - arranged in a circle around a central point.

Sources

Ahmadjian, V. "The Nature of Lichens." *Natural History* 91, no. 3 (1982): 31-34.

Beattie, A. J. "Ant plantation." *Natural History*, (1990): 10-14.

Billington, C. *Shrubs of Michigan*. Bloomfield Hills, NY: Cranbrook Institute of Science, 1943.

Blanchan, N. *Nature's Garden*. New York: Doubleday, Page and Co., 1901.

Burr, C. Wisconsin's Carnivorous Plants. *Wisconsin Natural Resources* 5, no. 3 (1981): 28-31.

Clark, J. S. "The Forest Is for Burning." *Natural History*, (1989): 50-52.

Cook, R. E. "Disturbing Plants." *Natural History* 90, no. 3 (1981): 22-26.

Cook, R. E. "Reproduction by Duplication." *Natural History* 89, no. 3 (1980): 88-93.

Curtis, J. *Vegetation of Wisconsin*. Madison, WI: University of Wisconsin Press, 1959.

Dana, W. S. *How to Know the Wild Flowers*. Boston: Houghton Mifflin, 1893.

Daniel, G., and J. Sullivan. *A Sierra Club Naturalist's Guide to the North Woods of Michigan, Wisconsin, and Minnesota*. San Francisco: Sierra Club Books, 1981.

Davidson, D. W. "Competition for Dispersal in Ant-Dispersed Plants." *Science* 213 (1981): 1259-61.

Densmore, F. *How Indians Use Wild Plants for Food, Medicine and Crafts*. New York: Dover, 1974.

Durant, M. *Who Named the Daisy? Who Named the Rose?* New York: Congdon and Weed, 1976.

Eastman, J. "The Ghost Forest." *Natural History* 95, no. 1 (1986): 10-16.

Eckstein, R. "Eastern Hemlock in North Central Wisconsin." DNR Research Report 104 (1980): 1-10.

Eggers, S., and D. Reed. *Wetland Plants and Plant Communities of Minnesota and Wisconsin*. St. Paul, MN: US Army Corps of Engineers, 1986.

Elias, T. S. *Trees of North America*. New York: Van Nostrand Reinhold Company, 1980.

Fassett, N. *Spring Flora of Wisconsin*. Madison, WI: University of Wisconsin Press, 1967.

Heat-Moon, W. L. "A Working Tree." *Outside* 16, no. 10 (1991): 109-14.

Houk, R. *Eastern Wildflowers*. San Francisco: Chronicle Books, 1989.

Janke, R. *The Wildflowers of Isle Royale National Park*. Houghton, MI: Isle Royale Natural History Association, 1962.

Jones, L. R., and F. V. Rand. *The Handbook of Vermont Shrubs and Woody Vines*. Rutland, VT: Charles E. Tuttle Company, 1909.

Kerasote, T. "The Joy of Alder." *Audubon* 87, no. 4 (1985): 10-13.

Klimas, J. E. *Wild Flowers of New Hampshire and Vermont*. New York: Walker and Company, 1968.

Kotar, J., J. Kovach, and C. Locey. *Field Guide to Forest Habitat Types of Northern Wisconsin*. Madison, WI: University of Wisconsin-Madison, 1988.

Kricher, J., and G. Morrison. *A Field Guide to Eastern Forests*. Boston: Houghton Mifflin, 1988.

Laine, W. G. "Wisconsin's Milkweeds." *Wisconsin Natural Resources* 19, no. 5 (1985): 19-22.

Lanner, R. M. *Autumn Leaves*. Minocqua, WI: NorthWord Press, 1990.

Lawrence, G. *A Field Guide to the Familiar*. Englewood Cliffs, NJ: Prentice Hall, 1984.

Linton, J. "A Tension Zone of Our Own." Wisconsin Natural Resources 11, no. 11 (1987): 6-7.

Lyons, J., and S. Jordan. *Walking the Wetlands*. New York: John Wiley and Sons, 1989.

Martin, A., H. Zim, and A. Nelson. *American Wildlife and Plants*. New York: Dover Publications, 1951.

Mathews, F. S. *Field Book of American Wild Flowers*. New York: G. P. Putnam's Sons, 1902.

Meyer, T. "Wild Beauties." *Wisconsin Natural Resources* 14, no. 3 (1990): 17-20.

Morley, T. *Spring Flora of Minnesota*. Minneapolis: University of Minnesota Press, 1969.

Newcomb, L. *Newcomb's Wildflower Guide*. Boston: Little, Brown and Company, 1977.

Palmer, E. L., and H. S. Fowler. *Fieldbook of Natural History*. New York: McGraw-Hill, 1949.

Peattie, D. C. *A Natural History of Trees*. Boston: Houghton Mifflin, 1948.

Peterson, L. E. *A Field Guide to Edible Wild Plants*. Boston: Houghton Mifflin, 1977.

Pielou, E. C. *The World of Northern Evergreens*. Ithaca, NY: Comstock Publishing Associates, 1988.

Roth, C. E. *The Plant Observer's Handbook*. Englewood Cliffs, NJ: Prentice-Hall, 1984.

Smith, H. *Michigan Wildflowers*. Bloomfield Hills, MI: Cranbrook Institute of Science, 1961.

Stokes, D. W. and L. Q. *A Guide to Enjoying Wildflowers*. Boston: Little, Brown and Company, 1985.

Stokes, D. W. and L. Q. *The Natural History of Wild Shrubs and Vines*. New York: Harper and Row, 1981.

Van Stappen, M. "Leopold's Last Stand; Hard Times for North Woods Hemlocks." *Tuebor Terra*, (1990): 28-31, 42-43.

Voss, E. G. *Michigan Flora*, vol. 1. Bloomfield Hills, MI: Cranbrook Institute of Science and University of Michigan Herbarium, 1972.

Voss, E. G. *Michigan Flora*, vol. 2. Ann Arbor, MI: Cranbrook Institute of Science and University of Michigan Herbarium, 1985.

Weiner, M. A. *Earth Medicine Earth Food*. New York: Fawcett Columbine, 1972.

Pfeifer-Hamilton Publishers produces quality gift books celebrating the special beauty and unique life-style of the north country.

by Shawn Perich
Fly-Fishing the North Country
Fishing Lake Superior
The North Shore

by Scott Anderson
Distant Fires

by Nadine and Craig Blacklock
Gooseberry

by Bob Cary
Root Beer Lady

by Sam Cook
Up North
Quiet Magic
CampSights

by Laura Erickson
For the Birds

by Michael Furtman
Canoe Country Camping

by Mark Stensaas
Canoe Country Wildlife

by Jerry Wilber
Wit and Wisdom of the Great Outdoors

by Douglas Wood
Paddle Whispers

Call us at 800-247-6789 for a complete catalog.

Pfeifer-Hamilton Publishers
210 West Michigan Duluth MN 55802-1908